Rochester Roundabout
The Story of High Energy Physics

Rochester Roundabout
The Story of High Energy Physics

John Polkinghorne, FRS

Longman

Longman Scientific & Technical
Longman Group UK Limited
Longman House, Burnt Mill, Harlow
Essex CM20 2JE, England
and Associated Companies throughout the world.

First Published in 1989

British Library Cataloguing in Publication Data
Polkinghorne, J.C. (John Charlton), *1930–*
   Rochester roundabout.
   1. High energy physics
   I. Title
   539.7'2

ISBN 0-582-05011-1

Set in 10½/13 pt Times Roman
Produced by Longman Singapore Publishers (Pte) Ltd.
Printed in Singapore

Contents

To my former colleagues in the international community
of high-energy physicists.

Preface

If we are truly to understand military history it would not be enough simply to read the despatches of the generals. The reminiscences of those in the ranks would also be needed to shed a complementary light on what was happening. In that understanding, I have written the story of high-energy physics during the period in which I saw service as a theorist. As the set-piece engagements of that long campaign I have selected the 'Rochester' Conferences, international state-of-the-art reviews, occurring initially yearly and later at two-year intervals. Their Proceedings, particularly in the earlier years, provide a more informal, and consequently more realistic, account of how physicists think than can be obtained by perusal of the pages of the *Physical Review*. I have, of course, also reread other versions of the story. I am particularly indebted to 'Bram Pais's *Inward Bound* for its valuable chronological table, which jogs the memory of what happened when.

Pais describes his own account as a 'memoir', quoting one of the definitions offered by the *Oxford English Dictionary*: 'A record of events, not purporting to be a complete history, but treating of such matters as come within the personal knowledge of the writer, or are obtained from certain particular sources of information.' I offer my own tale in a similar spirit. It is necessarily written from a particular perspective. An artilleryman sees the battle differently from an infantryman. While I endeavour to acknowledge appropriately the immense importance of experimental ingenuity and persistence in the development of high-energy physics, it is a theorist's eye view which is given here. Even among the infantry, different regiments have different traditions, and a careful reader will perceive a variety of differences of emphasis, and even of subject, between Pais's account and my own. Lastly, a soldier may be pardoned if he accords special interest to those skirmishes in which he actually took part. I allow myself, from time to time, the indulgence of commenting a little on those problems that engaged my own attention.

Each 'Rochester' Conference is given its own chapter. In addition, the first chapter introduces the state of the art in high energy physics as it existed in 1950 and the last chapter uses the preceding story as a test case for the evaluation of ideas and claims in the philosophy of science. Since it would be a trifle ludicrous to speak about Conference N in Chapter $(N+1)$, I have taken a leaf out of the mathematicians' book and started counting at zero.

One final word. If there is (as I hope there is) a certain lightness of tone in this book, do not let it deceive you about the seriousness of its intent. I am attempting to share something of the pleasure given by one of the most sustained and successful intellectual enterprises of this century. I am also, in my final chapter, endeavouring to defend, in the light of actual experience, a critically realist account of science which sees that enterprise as resulting in a verisimilitudinous knowledge of the structure of the physical world. Quarks and gluons are not useful and amusing manners of speaking; they are, at the appropriate level, the actual entities of which matter is composed.

I have dedicated this book to my former colleagues in the community of high-energy physicists, from whom I received so much, both intellectually and personally, over many years. I would like particularly to mention Nick Kemmer, who first aroused my interest in the use of mathematics to understand the physical world; Abdus Salam, who guided my doctoral research; Murray Gell-Mann, with whom I spent a formative and highly enlightening postdoctoral year; the late Lev Landau, whose deep insight stirred my mind, as it did the minds of many others; and Peter Landshoff, whose skill in the collaborative work we did together enabled me to do much more physics than I could ever have accomplished on my own. I also pay tribute to my other collaborators and to a succession of bright students whose initial researches I had the privilege of guiding.

I am grateful to Professor Maurice Jacob and Professor John C. Taylor for reading the manuscript and making helpful comments. The responsibility for its final form is, of course, mine alone. I am also grateful to Professor Jacob for hospitality at the Theory Division of CERN which enabled me to make use of that institution's admirable library. I thank my wife Ruth for help in correcting the proofs. It has been a great pleasure for me once again to work with Dr Michael Rodgers. I am grateful to him and the editorial staff of Longman for their assistance in preparing the manuscript for press.

John Polkinghorne
Trinity Hall,
Cambridge, 1988

List of Illustrations

Nomenclature

In the account that follows I shall use certain items of nomenclature taken from the common stock of high-energy physics. We encounter four types of basic force in elementary particle physics: strong (e.g. nuclear forces), electromagnetic, weak (e.g. responsible for beta decays) and gravitational. Particles which participate in the strong force are called *hadrons*. More specifically, if they have half-odd-integer spin ($\frac{1}{2}$, 3/2, ...) they are called *baryons*; if they have integral spin (0, 1, ...) they are called *mesons*. Baryons of non-zero strangeness are called *hyperons*. Particles of spin $\frac{1}{2}$ which do not have strong interactions are called *leptons*.

In numerical assessments we use *order of magnitude* to denote a power of 10. If something is larger by two orders of magnitude it is 100 times bigger.

Energies are measured in Mev (the energy acquired by an electron falling through an electromagnetic potential of a million volts) or Gev (= 1000 Mev).

The symbol α is used to denote the fine-structure constant, a natural measure of the strength of electromagnetism; its value is approximately 1/137.

Frequently, I and J are used to denote total isospin and angular momentum, respectively (see glossary for definitions of these quantities).

Tomorrow is going to be wonderful, because tonight I do not understand anything.

Niels Bohr

0   *Pre-Rochester*

The pursuit of pure physics largely lapsed during the 1939—45 war, in favour
of the application of science to the war effort. When hostilities ceased it was
time to return to fundamental questions about the pattern and structure of
the physical world, issues whose elucidation has always presented an irresistible
challenge to people of the highest intellectual talent. It proved to be a period
of great opportunity.

    There was much goodwill towards the physics community on the part
of those who had control of the allocation of resources. In the immediate
aftermath of victory, the construction of the atomic bomb did not seem as
morally ambiguous or as globally dangerous an act as later it was to appear.
Radar had been another development which showed clearly the power of
physics to solve problems of great practical significance. Luis Alvarez, one
of the leading figures in postwar high energy experiments, said: 'Right after
the war we had a blank cheque from the military because we had been so
successful . . . we never had to worry about money.'[1] That euphoric phase
was not to last and the question of funding has been a chronic worry for high-
energy physics for many years. To learn about the constituents of matter we
must investigate what is happening on the shortest possible length scales, much
shorter than one fermi (10^{-13} cm). The only access available to such minute
phenomena is through collisions taking place at very high energies. The shorter
the distance to be probed, the higher the energy which is required in order
to do so. One reason is that penetrating power costs energy. The deeper one
wishes to look inside a proton, the faster one's projectile must be moving
to get that far. Another reason, yet more fundamental, is provided by quantum
theory. Its irreducible 'fuzziness' means that the distance sampled varies
inversely with the momentum of the sampling particle. (Technically: in terms
of wave/particle duality one needs short wavelengths, and that means high
momentum.[2])High energies cost money, for they can only be produced from
accelerating machines built by employing precision engineering on a vast scale.

The contemporary generation of such machines are kilometre-sized, cost many hundreds of millions of pounds to construct and are capable of looking for structure on a scale of less than 10^{-16} cm. It is only by building big that one can think small.

That the rich, developed nations of the world should feel able to spend some tiny fraction of their GDP on the investigation of the fundamental structure of the physical world is a test of their degree of civilisation. The American accelerator builder, R.R. Wilson, testifying before a Congressional Committee about a proposal for a new machine, is said to have answered the question, 'What will this project do for the defence of the United States?' with the admirable reply, 'Nothing − but it will help to make the United States worth defending.' It was my job, for a while at the end of the seventies, to ask the British taxpayer, through the Science Research Council, for the forty million pounds or so that were then being spent each year on research in high energy physics by the United Kingdom. Naturally, the request was subject to careful scrutiny by my fellow scientists in other disciplines, who were also seeking funds from the same source. In defence of my proposals I could cite various spin-off effects from the work of the high-energy physics community − clever technological tricks with potential applications elsewhere; the development of a pool of trained and skilful scientists, and so on. But in the end, the only totally honest and totally adequate reason was that it was work intrinsically worth doing. To find out of what stuff the infinite variety of the world is composed is one of the grandest enterprises of humanity.[3] The tale of this book is of thirty years of such endeavour. I believe that the remarkable discoveries that resulted are sufficient warrant for the expenditure of so much talent, time and treasure in the exercise.

As physicists dispersed from great wartime collaborations, such as the Manhattan project, they took with them more than simply the governmental goodwill that their efforts had earnt. There had been important technical developments (for example, in the use of microwaves) which would speedily find application in high-energy physics. There was also the experience of the power of collaborative effort. General Groves and Robert Oppenheimer had assembled on the Los Alamos mesa in New Mexico perhaps the greatest concentration of scientific talent ever brought together in one place for a common purpose. Problems of enormous complexity and difficulty had been solved by their joint efforts. When one reads reminiscences of those days one consistently receives the impression of a community completely entranced by what Oppenheimer called the 'sweetness' of the scientific challenge. For the subsequent development of high-energy physics it was a timely experience. The heroic primeval days of Ernest Rutherford and his lab assistant, Mr Crowe, doing a fundamental experiment on a laboratory bench, lay long in the past.

The future would involve teams of specialists, united in the construction and exploitation of some large facility.

Three contributions must blend in order to make progress in elementary particle physics. There must be a source of the phenomena, a means for their investigation, and a theoretical point of view from which to attempt their understanding. In 1945 all three were available and they would develop rapidly in the years following.

Some phenomena occur without their having to be artificially contrived. In the early years of the twentieth century much of the progress in probing the structure of matter had resulted from exploiting naturally occurring radioactivity. By the late nineteen thirties the investigation of the structure of nuclei had become a separate subject (low-energy nuclear physics) largely distinct from the investigation of the elementary particles composing all matter (high-energy physics). For the latter, the only facility provided free by nature was the shower of cosmic rays bombarding the Earth from outer space. These rays result in energetic particles whose interactions can be studied. There are difficulties, however. You have no control over what happens and you just have to take what comes. The most energetic particles are found at the highest altitudes, for cosmic rays lose energy as they traverse the atmosphere. Thus mountain tops and balloons figure largely in the story. Despite these problems, in the immediate postwar years much information of great significance was derived from cosmic ray studies. Yet the eventual accumulation of detailed and accurate information required the development of artificially produced beams of particles, with all the definition and control that that could offer. Today, cosmic rays continue to provide a window looking on to ultra-high-energy events, but it is glass through which we are only able to peer darkly.

The earliest particle accelerator was the cathode ray tube, in which an applied voltage draws electrons from a heated source. It was the means by which the first elementary particle was discovered, when J.J. Thomson identified cathode rays as beams of electrons in 1897. The device exemplified two requirements, basic to all subsequent accelerators: a vacuum, so that the accelerated particles are not jostled out of orbit by intervening matter, however tenuous, and a way of feeding in energy to produce the acceleration. This energy is always conveyed electromagnetically. Modern accelerators come in two shapes: straight and circular. (A theoretical friend of mine, bored by being shown round such facilities, once said that he would only be interested in seeing another if it were square!) In essence, the linear machines operate by causing particles to 'fall' through a strong electric potential, gathering momentum as they do so. The prototype of such devices was constructed in Cambridge by Cockcroft and Walton in 1932. It was the means of producing

3

the first artificially induced nuclear reaction. Cockcroft and Walton's machine was vertical, but as energies and sizes grew, a horizontal orientation became necessary. Modern linear accelerators are kilometres long.

The first-generation circular machines were the cyclotrons. They used an electric field to give the particles a series of 'kicks' to enhance their energy, whilst employing a magnetic field to constrain them to travel in circular orbits. The radius of the orbit increases with the particle's energy but, provided it is not moving so fast as to make relativistic effects significant, the particle takes the same time to traverse the orbit, whatever its radius. This regularity is of tremendous help, for it makes it possible to give all the particles the maximally effective kick each time, in whatever orbit they are travelling. The principle of the cyclotron was invented in Berkeley by Ernest Lawrence in 1929. Crude models were constructed in the following spring and the first physics results produced in 1932, a few months later than the work of Cockcroft and Walton. The first cyclotron was four inches wide, the largest ever constructed had a diameter of sixty inches.

Progress beyond that point needed innovation to cope with the effects due to the relativistic increase of mass with energy. This required a delicate tuning of the rate at which kicks were delivered to ensure that they continued to be maximally effective. This modulation of the radio-frequency in the accelerating cavity is the basis of the synchrocyclotron. It is possible as a practical procedure because of a principle called 'phase stability', independently recognised by the Soviet physicist V.I. Veksler in 1944 and by the American Ed McMillan in 1945. One is attempting to steer the particles along the right track at the right rate. The contrariness of real life means that one's efforts will never be precisely correct. The question is whether the unavoidable margin of error will lead to ever-increasing wobbliness or will it be damped out to give a convergent stability. The answer was that there is an intrinsic negative feedback in the system which works for orderly acceleration. If a particle is moving slightly too fast it will receive a weaker kick which will bring it into line by slowing it down for the next cycle, whilst if it is moving too slowly its kick will be greater, helping it to catch up. Therefore a bunch of particles can be accelerated successfully and they will stay together without being dispersed and lost. These machines produce particles in pulses, as a bunch of particles spirals through the accelerating process. In 1946 a synchrocyclotron which was 184 inches in diameter came into operation at Berkeley. It was to be the biggest machine of this kind ever built.

The trouble with any kind of cyclotron was that the particles followed orbits of increasing radius. To provide the necessary magnetic field as they spiralled out one had to construct huge disc magnets which would cover their path. These magnets used vast quantities of expensive high-grade steel. It

4

would obviously be a great saving if the particles could be constrained to move in circular orbits of unchanging radius, whatever their energy. This is the principle of the synchrotron. A bunch of particles is accelerated and as their energy increases the magnetic field increases in step so that the radius of the orbit remains constant. It is then possible to use a circular array of magnets rather than a big disc. Clearly, the successful operation of a synchrotron calls for very subtle controls and stabilities. McMillan constructed his first synchrotron at Berkeley in 1949. Modern synchrotrons are kilometres in diameter and can have a thousand electromagnets around their perimeter, each positioned to an accuracy of a tenth of a millimetre.

Accelerator improvement has played an essential enabling role in the development of high-energy physics. The first machines were built by those who were themselves to use them for experiments, but as sophistication increased there arose an army of accelerator physicists and engineers, specialising in synchrotron design and building. In the period surveyed by this book, machine planning and construction became an increasingly separated activity. We shall not be able to do it justice in this theorist's eye account but will simply notice two important advances which occurred. The first was the development of the alternating-gradient synchrotron. It used a clever way of producing what is called strong focussing, an invaluable aid to stability. This was also economically effective for it resulted in more tightly contained beams which could be controlled by smaller-sized magnets.

The second development related to how the particle collisions, which are the basis of the experiments, are brought about. All the devices described so far result in what are called 'fixed target' collisions. A beam of high-energy particles from the accelerator impinges upon a block of material, or a container of (say) liquid hydrogen. The advantage of this is that the density of the material in the target helps to produce many events (collisions). There are a lot of particles there waiting to be hit. The disadvantage is that relativistically it is very inefficient to strike a stationary target. The effective energy available in such 'sitting-duck' experiments only increases with the square root of the beam energy. Thus to investigate a phenomenon at twice as high an energy (or equivalently to probe on a length scale half as small again) you have to increase your accelerator energy four times. Much more effective would be to allow two beams to collide with each other. Then doubling the accelerator energy also doubles the effective collision energy.[4] Great! you say, but there is a snag. The beams are really very tenuous and they almost always pass through each other without any collisions taking place. (It is like crossing the road in a country lane rather than launching out into the traffic of Piccadilly.) Thus, if we are to perform experiments in this energy-effective, colliding way it is necessary to find some way to thicken up the beams.

The answer lies in storage rings, where many beam pulses can be accumulated over periods of hours and then eventually allowed to impinge upon each other when enough particles are present to ensure that something happens at the beam intersections. One can either have two rings which can be made to interact tangentially with each other or, alternatively, particle and antiparticle beams can circulate in opposite directions in the same ring and be made to cross each other when required. Clearly, the building of storage rings poses formidable stability problems, since the beams they contain must be capable of circulating for long periods without appreciable particle loss. One absolute essential for their successful construction was the attainment of very high quality vacua. Colliders first came into operation in the middle sixties. They have a very important role in high-energy physics today. CERN is currently constructing a ring to store electrons and positrons (LEP — the Large Electron—Positron ring) which will be 10 kilometres in diameter, almost burrowing under Geneva airport on one side and under the Jura mountains on the other. Of course, all colliders need a synchrotron to feed them with the particles they store. Their very high energy is purchased at the cost of a lack of flexibility. The particles extracted from a synchrotron can be made to collide with a variety of fixed targets, and some of these collisions can themselves yield as consequences secondary beams of other types of particles to use in further experiments. There is a great variety of things that one can do. At a colliding-beam facility only one basic experiment takes place, the interaction of the beams. One social effect of this narrowing of opportunity has been to concentrate yet further the activities of high-energy experimentalists into large collaborations, involving literally hundreds of physicists. There is work for all of them because the events taking place at very high energies are very complex and they need extensive instrumentation and analysis for their full investigation.

The second provision necessary for successful elementary particle physics is some form of apparatus to detect the results of particle collisions. All detectors depend upon electromagnetic effects to make perceptible in the laboratory the presence of microscopic particles which otherwise would be unobservable. It is, therefore, only charged particles which can be made visible in this way. The presence of neutrals has to be inferred in some less direct manner.

In 1911, when Rutherford made the fundamental discovery that the atom had a nucleus within it, he did so by noting flashes made by the scattered alpha particles as they struck a zinc sulphide screen.[5] Other detecting devices in use early in this century were the Geiger counter and the cloud chamber. Both depended on the ionising effects of moving charged particles. In the cloud chamber this produced a condensed trail of little water droplets as the

6

particle traversed the saturated water vapour filling the chamber. (The device was invented by C.T.R. Wilson, who was a meteorologist with a specialist interest in Scotch mist.) In the Geiger counter the passage of the charged particle discharged a high voltage and the resulting crackle was amplified to produce an audible click to register the particle's passing.

In all these cases a microscopic trigger produces a macroscopic signal. In the period that this book surveys, the detectors in use fell into three families, corresponding to the three prototypes already described. The zinc sulphide screen developed into scintillation and Čerenkov counters, producing pulses of light, nowadays scanned by photomultiplier tubes rather than by a red-eyed lab assistant. The successor of the cloud chamber has been the bubble chamber, in which the track is produced by a trail of bubbles formed in a superheated liquid by the ionizing effect of the particle traversing it. Bubble chambers (first used in 1953) are superior to cloud chambers because they can be reused more quickly, they give a finer definition of the path, and the denser material they contain induces more interactions. Another form of visible-track detector, much used in the earlier years of our period, was a stack of photographic emulsion. The ionising particles leave an image which can be developed, but the analysis of the resulting imprints calls for much painstaking labour. The descendants of the Geiger counter are devices depending on various forms of electrical discharges between plates or wires maintained at high voltage. This is the basis of detectors such as spark chambers or proportional wire chambers.

As with the accelerators, our account can only be a summary. The great development in the period under review was the increasing size and sophistication of these devices and the ever more extensive use of electronic monitoring to trigger and record their operation. Such use of electronic techniques enables the experimentalist to concentrate his search and to seek and analyse rare events, recognised by characteristics specified beforehand. This ability to focus on needles in haystacks is of very great importance in high-energy collisions where many things happen but those of greatest interest are often of a corresponding rarity. The method carries with it, of course, the danger of overlooking the unexpected, which has not been programmed into the search and selection procedure.

The third desideratum for elementary particle physics is a theoretical setting for interpretation and understanding. One must have at least some tentative knowledge of the nature of the forces with which one is dealing. Electromagnetism has been studied for a hundred and fifty years or more. The character of the theory was transformed in the twentieth century by the discovery of the quantum mechanical wave/particle duality of the photon. Gravity has been known and studied for even longer but its extreme weakness in ordinary subatomic circumstances − it is thirty-seven orders of magnitude weaker than electromagnetism − meant that it was largely neglected by elementary particle physicists during the period under review. It is only latterly, with growing interest in the convergence of the fundamental sciences of the very large (cosmology) and the very small (particle physics) in the very high-energy regimes of the early universe, that gravity has moved on to centre stage.

Two sorts of nuclear forces were also known: the strong nuclear force (a hundred times more potent than electromagnetism) which holds nuclei together, and the much weaker nuclear force which causes some nuclei to undergo beta decay and fall apart. Important progress had been made in the 1930s in the understanding of these two forces. In 1934 it occurred to the Japanese physicist Hideki Yukawa that, just as the photon is the carrier of electromagnetic forces, so there might also be a particle which mediated the strong nuclear force. The latter's known short range could then be attributed to this particle's having a mass, in contrast to the massless photon which was responsible for the long-range of electromagnetic forces. Moreover, there was a natural connection between the range of the nuclear force and the corresponding mass expected to be possessed by Yukawa's hypothetical particle. This turned to be about two hundred times the mass of the electron, a value about a tenth of that of the proton. This intermediate character led to the particle's being dubbed the mesotron, soon abbreviated to meson. It seemed at the time a striking confirmation of this idea when a few months after Yukawa's suggestion a particle of just such a mass was identified in cosmic rays.

Further progress was made in understanding the nature of the strong nuclear force when its charge-independent character was recognised. This property refers to the fact that the force appeared to be the same whether it was operating between two protons or two neutrons or a proton and a neutron. However much these two constituents of nuclei might differ in their electromagnetic characters, when it came to the strong nuclear force they appeared to be two peas in a pod in similarity. This emboldened Heisenberg to introduce the generic term 'nucleon' to refer to them both. Given that people believed they had reason to think that there were charged mesons, it was a non-trivial problem to work out nevertheless how their exchange could lead to totally charge-independent forces. This conundrum was brilliantly solved in 1938 by Nicholas Kemmer. It was of extraordinarily great importance for later developments how he did so. Heisenberg had made a formal analogy between the two-state system of the nucleons (proton, neutron) and the familiar two-state system which quantum mechanics assigns to the lowest state of non-zero angular momentum corresponding to the two spin-½ states of spin up

8

and spin down.[6] If an interaction is rotationally symmetric in space (that is to say, if there are no special preferred axes, so that one direction is as good as another) then the way a spin-½ particle behaves is independent of its actual spin value. Kemmer's trick turned out to treat, in an exactly analogous way, the interaction of protons and neutrons with mesons in a 'rotationally' symmetric fashion. This then guarantees that proton and neutron will have identical, charge-independent interactions. Of course, 'rotation' has to appear in inverted commas because we are concerned with transformations changing a proton into a neutron, not with pirouetting in the laboratory. These transformations could be expressed from a formal mathematical point of view as if they were 'rotations' in a fictitious 'space' − a space associated with a particle's internal characteristics (whether it is a proton or a neutron) and not with external behaviour in the world of the laboratory. Because mathematically these transformations look the same as rotations, the word 'spin' got associated with the notion. In a colossal terminological blunder the idea actually got called 'isotopic spin'. Since substitutimg a proton for a neutron changes a nucleus into one of its isobars, and not one of its isotopes, the name should have been 'isobaric spin'. Fortunately, in contemporary terminology it has got abbreviated to the innocuous 'isospin'. By whatever name, the notion of an internal symmetry, linking together different sorts of particles, has proved to be of very great importance. Kemmer found that he needed three kinds of mesons (respectively, electrically positive, neutral and negative). That was due to the mesons being associated with isospin 1, a fact that was not fully recognised and exploited until after the war.

The 1930s had also seen important progress in understanding the weak nuclear force. The most significant development was the suggestion by Pauli in 1930 that difficulties about energy conservation, which appeared to arise in these interactions, could be resolved if the amount 'missing' were to be carried off by a hypothetical new particle, electrically neutral and of zero mass, which eventually was given the name neutrino.[7] Neutrinos only possess this very weak form of interaction with other matter and so they can pass easily through the Earth, which is transparent to them. It was not until 1956 that a neutrino was detected directly. In 1933 Enrico Fermi proposed his famous theory of beta-decay processes. It pictures them as involving the simultaneous point interaction of four particles. In the decay of the neutron, for example, these would be the neutron itself and the proton, electron and antineutrino into which it is transformed.

So far we have been considering specific theoretical questions: the nature of the fundamental interactions and of the particles which participate in them. This discussion has to take place within an overall theoretical framework. The character of the latter is determined by those twin discoveries

9

which give the theoretical physics of the twentieth century its particular character, very different from that of the centuries which precede it. Because elementary particles are small, their proper description requires quantum theory. Because they move very fast, their proper description also requires special relativity. The combination of these two disciplines is a highly non-trivial synthesis, both to achieve and in its consequences. The whole is very much more than the sum of the two parts. The two most important ideas necessary for the carrying through of this programme were both contributed by Paul Dirac.[8] In 1927, by his application of quantum theory to electromagnetism, he created the first example of a quantum field theory. Here we find the perfect language in which we may speak of wave/particle duality without any danger of paradox. Because we are concerned with a field (something spread out over space and time) wavelike properties are present; because the characteristic effect of quantum theory is to make things discrete (that is to say, countable), energy comes in certain numbers of packets, immediately interpretable as certain numbers of particles. The synthesis of quantum theory and field theory is simultaneously the synthesis of wave and particle.

Dirac's other great achievement was his famous relativistic equation of the electron, published in 1928. It showed how particles of spin ½ should be described in relativistic quantum theory. It is a very well known story how certain difficulties with negative energies led Dirac eventually to a brilliant resolution through the recognition of the existence of antimatter. This insight received a prompt and triumphant vindication with the discovery in 1931 of the positron, the antiparticle brother of the electron, in cosmic rays.

The combination of relativity and quantum theory possesses properties not to be found in either separately. This theoretical richness found its first illustration when Pauli showed in 1934 that the observed relation between spin (a particle's intrinsic angular momentum) and statistics (how particles behave in aggregate) followed as a consequence of relativistic quantum theory.

Those prewar years had seen great advances in elementary particle physics. Yet when the clouds gathered in 1939, a great difficulty remained unresolved which threatened the consistency of the whole enterprise. It related to calculations with quantum electrodynamics, the theory of the interaction of photons and electrons. The parameter which measures the intrinsic strength of that interaction, the fine-structure constant α, is very small, its value being about 1/137. This enabled one to attempt to calculate by successive approximation, working in ascending powers of α. The first-order terms, proportional to α itself, made good sense of the experimental data. One did indeed seem to be on to something. It was natural to press on to gain the extra refinement expected to be afforded by calculating the corrections of second order in α. The result was disastrous. The square of α might be small enough, but when one computed the coefficient multiplying it the answer came out as infinity! This meant that the theory as it stood was, as a matter of fact, meaningless. This awful problem remained unresolved when world conflict put such anxieties out of people's minds for a while. The return of peace restored it to the agenda.

One important insight related to the problem had already been gained before the war. It referred to the curious properties of the quantum mechanical vacuum. The vacuum is the lowest energy state of a system. It is a peculiarity of quantum theory that when there is nothing there it does not mean that there is nothing happening. One can most readily understand how this counterintuitive behaviour comes about by considering a simple, but fundamental, physical system, the simple-harmonic oscillator. If you want a concrete example, think of a pendulum. In classical physics the lowest energy state of the pendulum (that is to say, its vacuum) is when the bob is hanging at the bottom at rest. When we introduce quantum theory, Heisenberg changes all that. He rejects such stillness, for his uncertainty principle will not allow us both to know where the bob is (at the bottom) and what it is doing (it is at rest). Instead a judicious compromise is called for — nearly at the bottom, nearly at rest, but not quite. There is an irreducible quiveriness in the quantum pendulum, which the experts call its 'zero-point motion'. Augmented to the complexity of a quantum field, this produces the quantum vacuum in all its inescapable restlessness, manifesting the continual coming-to-be and ceasing-to-be of transient blips of energy. A particle inserted into such a vacuum gains energy (or, equivalently, mass, since we all know that $E=mc^2$) from its interactions with these fluctuations. Those same fluctuations also make the vacuum behave like an electrically polarisable medium and that changes the charge of the particle also. Since the vacuum is everywhere, these mass and charge modifications are always present. By 1939 it had been realised that they were closely associated with the disastrous infinities.

On the 1st of June 1947, twenty-five people gathered at a physics conference on Shelter Island, just off the coast of Long Island, New York. Duncan McInnes of the Rockefeller Institute had had the idea of assembling a small group of experts to discuss the current state of play in elementary particle physics. The National Academy of Sciences had given $3100 to fund the occasion. Willis Lamb, of Columbia University, described how he had measured a small difference in atomic energy levels in hydrogen (now called the 'Lamb shift'). Atomic spectroscopy, from the time of Balmer onwards, has played a crucial role in the advance of fundamental physics. Not least, this has been so because its great accuracy has enabled it, time and again, to pose precise quantitative questions that theory must attempt to answer. Lamb was using his wartime expertise with microwaves to perform a particularly accurate and interesting measurement. The two levels that he was examining were degenerate (that is, they had coincident energies) according to calculations based on the Dirac equation. Usually in physics, degeneracies arise from some sort of symmetry consideration which requires two things to be equal. (If my body is perfectly symmetrical about a vertical axis I shall need precisely the same sized shoes for both my feet.) Sometimes, however, the degeneracy is just an accident of circumstance. (The near equality of the angular diameters of the Sun and Moon as viewed from the Earth is presumed to have that character.) In the nature of things, accidental degeneracies are more likely to turn out to be approximate rather than exact. Only if Dirac's equation were to be the absolute truth (rather than a very good approximation to the truth) would we expect to find those two levels at exactly the same energy. Lamb found that, in fact, there was a separation between them. There is nothing like a good experimental measurement needing explanation to concentrate the theoretical mind. Lamb had thrown down an experimental gauntlet which the theorists were quick to pick up. Within a few days of the Shelter Island Conference, Hans Bethe, with the back of an envelope and a little inspired cookery, had shown that a credible explanation of the Lamb shift lay in the vacuum effects which the Dirac equation did not take into account.

It was clearly necessary to replace inspired 'ball-park' estimates with sober calculational procedures. It was also known that the electron's magnetic properties (its magnetic moment) were not quite those predicted by the Dirac equation. Thus there was a pressing need to find a way to extract sensible finite results from the débâcle of prewar quantum electrodynamics. The key idea required lay in the precise identification of the source of the unwelcome infinities. They resided solely in the contributions made to the electron's mass and charge by the vacuum fluctuation effects already described. Eventually, a daring proposal was made. Since these effects are always present there is no way by which they can be separated out from other contributions to the mass and charge (the so-called 'bare' contributions not due to interactions). Therefore, at one stroke the infinities could be removed by simply equating all contributions to mass and charge (including the formally infinite ones) to the finite, actually observed, values of these physical constants. This procedure was called 'renormalisation'. The result was an immediately well-defined theory. It was also, from the start, a highly successful theory. Its first quantitative achievement came in December 1947 when Julian Schwinger calculated the correction to order α to the electron's magnetic moment (the anomalous moment), in excellent agreement with its measured value. Many subsequent developments have refined the accuracy of the experimental results, but the theory has always been able to keep pace with corresponding natural

12

refinements of calculation (in particular, working to higher powers in α) which have yielded corrections in excellent agreement with the observations. Richard Feynman, one of the great figures in postwar quantum electrodynamics, has described the theory as 'the jewel of physics — our proudest possession'. In a striking analogy he compares the accuracy it achieves today with a residual uncertainty of the width of a human hair in the distance between Los Angeles and New York.[9]

Renormalisation theory need not just be considered as a clever calculational trick. The infinities arise from effects taking place on very short length scales, shorter than we have been able to investigate experimentally. It is conceivable that novel effects intervene at these short distances, in ways not known to us, to produce the observed finite mass and charge.[10] Our ignorance does not allow us to calculate these effects, but we can compensate for that by feeding in the measured constants. Renormalisation is then seen as a device to produce moderate-energy knowledge despite ultra-high-energy ignorance.

So far, I have spoken rather cavalierly about the renormalisation programme. Its successful prosecution calls for very great skill and clarity of thought. Infinities are difficult and dangerous entities to handle mathematically. Unless great care is exercised one can easily get ridiculous results. Let me give a caricature example. Infinity (∞) is so big that doubling it does not make it any bigger. If I am infinitely tall, doubling my height will not make me any taller. Thus I am tempted to write

$$2 \times \infty = \infty,$$

from which, by subtracting ∞ from both sides, I get the spectacularly erroneous conclusion that $\infty = 0$. In quantum electrodynamic calculations it was necessary to keep very careful control of what was being done. An important guide was always to work covariantly (that is to say, to be careful to respect the requirements of special relativity at every stage) and to fulfil the constraints of gauge invariance (a rather technical but important symmetry possessed by electromagnetic interactions). Such prudence called for sophisticated calculational techniques for its implementation.

In his recollections of the Shelter Island Conference, Abraham Pais remembers 'late night talks on quantum electrodynamics with Schwinger, who was quiet throughout the sessions. Feynman trying to explain to me a new calculational method in field theory .... He would start by drawing some pictures'.[11] Schwinger and Feynman were to play very important roles in the new developments. Both were brilliant lecturers, but in contrasting styles. Schwinger, quiet in ordinary conversation, becomes like a man possessed on the platform. It seems to be the spirit of Macaulay which takes over, for he speaks in splendid periods, the carefully architected sentences rolling on, with every subordinate clause duly closing. It is a magnificent, polished, performance. Feynman, on the other hand, was the extrovert New Yorker. ebullient in expression, exuberant in gesture, demotic in vocabulary. There was sometimes something slightly tiring about the jocularity of the performance, but a profound insight into physics was vividly conveyed to the listener.

The difference in lecturing styles corresponded also to a difference in their ways of doing physics. Schwinger's work was very precise and elegant mathematically, but often hard to read. Feynman's work was much more picturesque and intuitive. He had his own way of going about quantum theory, based on his 'sum over histories' approach, whose basic notion is that an electron going from A to B does so by every possible path.[12] For Feynman, in a notorious and not always very helpful phrase, a positron was an electron 'going backwards in time'. Out of this richly idiosyncratic manner of thought emerged one of the most powerful tools for the expression and exploitation of relativistic quantum theory — Feynman diagrams. These were the pictures that Feynman kept on wanting to draw for Pais at Shelter Island. They portray the different ways in which interaction can take place through particle exchanges and they associate with each such process a quantum-mechanical contribution (technically, a probability amplitude) which is expressed in a form which we now call a Feynman integral. Feynman diagrams have about them an immediacy of insight and a flexibility of use which have made them an essential part of the vocabulary of modern thinking in theoretical physics. In a reminiscence of those early days Julian Schwinger has commented, 'Like the silicon chip of more recent years, the Feynman diagram was bringing computation to the masses.'[13] In fact, later on, he himself produced a 'theory of sources' which to the simple-minded looked remarkably like Feynman diagrams.

Shelter Island was followed in 1948 by a similar conference at Pocono Manor in Pennsylvania. Schwinger was able to present a largely completed account of his work on that occasion. Feynman's calculations were still in a less fully fashioned state. His time for full exposition was to come at the last of this sequence of three conferences (whose programmes were largely orchestrated by Robert Oppenheimer) which was held at Oldstone-on-Hudson in 1949.

So far we have been speaking of the American scene. Parallel but unnoticed developments had been taking place elsewhere. Sin-itiro Tomonaga, working in Japan in 1943 and writing in Japanese, had produced a 'many-time' formalism, essentially equivalent to Schwinger's later 'interaction representation'. Tomonaga went on to do renormalisation-type calculations.

14

This work only became known in the West in 1948, when the Japanese physicist wrote out of the blue to Oppenheimer. Stueckelberg, a brilliant but eccentric Swiss theorist working in Geneva, had produced some results similar to Feynman integrals. He published in the rather obscure journal *Helvetica Physica Acta* and so his work, too, attracted little attention. Eventually, Feynman, Schwinger and Tomonaga shared a Nobel prize for their fundamental contributions to the development of quantum electrodynamics.

The approaches of Feynman and Schwinger were so different in appearance that it was not immediately obvious that they were both doing the same thing. The question of reconciling their methods was not quite as important as had been that of establishing the equivalence of Heisenberg's matrix mechanics and Schrödinger's wave mechanics in the early days of quantum theory, but it was nevertheless a significant issue to address. The answer was found by Freeman Dyson, a young Englishman who had turned from pure mathematics to theoretical physics. Nick Kemmer tells a story of walking down King's Parade in Cambridge with Dyson and an Indian, Harishchandra. The latter had been working in theoretical physics but the confusions then present in the subject in the immediate postwar period had disillusioned him and he explained that he was moving over into pure mathematics (in which he subsequently had a distinguished career) because 'physics is in such a mess'. 'That's funny,' said Dyson, 'that's why I've decided to take it up.' He certainly did so with great effect, first moving to Cornell to learn the subject in a more 'dirty-handed' environment than was provided by the somewhat rareified atmosphere of the Cambridge of his day. Dyson made his mark by demonstrating the relation between Schwinger's ideas and those of Feynman, whilst at the same time making both more readily accessible to the ordinary student of the subject. I think that Schwinger makes oblique and ironic reference to these advances when he says that in 1949, 'There were visions at large being proclaimed in a manner somewhat akin to that of the Apostles who used Greek logic to bring the Hebrew God to the Gentiles.'[14] (Like so many talented theoretical physicists, Feynman and Schwinger were both Jewish.)

As a result of these developments, a systematic programme of investigation into renormalisation theory got under way. The central problem lay in the fact that as one moved to consider Feynman diagrams of higher orders in α, and consequently of greater complexity of structure, the infinities present tangled with each other in ways which were difficult to unravel. The general solution of the problem of these 'overlapping divergencies' was first given by a young Pakistani who had been working in Cambridge: Abdus Salam. His name will recur in our story. The clever but mercurial John Ward of Oxford soon produced a highly ingenious way of tackling the same problem, which he published in short papers characterised by compactness, difficulty and insight.

Willis Lamb's measurement of his level shift was not the only important contribution made to the Shelter Island Conference. The meeting also considered a problem which had been beginning to give anxiety before the war and which came into ineluctable prominence after it. The 'meson' discovered in cosmic rays was failing to behave in the way that Yukawa prescribed. Because it was supposed to be the carrier of the strong nuclear force, it would be expected to interact powerfully with matter. In particular, the negative mesons should be drawn towards nuclei by electromagnetic attraction between opposite charges and then quickly absorbed by the nuclear interaction once they came within its range. Cosmic ray studies in 1946 had shown, on the contrary, that the meson's interaction with matter was some ten to twelve orders of magnitude weaker than the Yukawa theory would have led one to expect. Was this just another case of a good proposal found eventually to be destined for the wastepaper basket? A young theorist present at Shelter Island, Bob Marshak, thought not. He proposed a way to salvage the idea. He did so by suggesting a two-meson hypothesis. There were, indeed, Yukawa mesons, the particles which we now call the pions. They were produced in cosmic rays but they were unstable and did not exist for very long. Before they could reach the cosmic-ray detectors they had decayed into a neutrino and another 'meson'. It was the latter which had been discovered before the war. According to modern terminology it was not a meson at all, for the latter term is reserved for particles with strong interactions. This particle's most powerful force was electromagnetic and so its non-absorption by nuclear matter was easily understood. The invention of a new particle to solve a difficulty was at that time a rare and risky undertaking. (The neutrino and the positron had been earlier successful examples.) Marshak recalls nevertheless that at Shelter Island, 'My recollection is that the two meson hypothesis was well received.' His act of theoretical boldness paid off. In fact, unknown to Marshak, one of the issues of the British journal _Nature_ had already published evidence for what we would now interpret as a pion decay. It was obtained by C.F. Powell's group in Bristol by extremely skilful investigation of cosmic rays using photographic emulsions flown in balloons. Marshak was equally unaware that a similar two-meson hypothesis had been proposed earlier by several Japanese theorists. As with the work of Tomonaga, their papers, published in wartime and in Japanese, were completely unknown in the West. When the paper formally putting the two-meson hypothesis forward was published, its authors were Bethe and Marshak. The latter tells us that, 'I decided to enlist Bethe's help in writing the paper because of his knowledge of cosmic ray data.'[15]

Satisfactory as this resolution was, it replaced one perplexity by another. Despite its name, μ-meson (nowadays abbreviated to muon), the particle discovered in cosmic rays was not really a meson but rather a lepton, a heavy brother to the electron. Its interactions were identical to those of the latter, only its mass was different. It was far from clear why nature should have chosen to repeat itself in this otiose fashion. It is only in the course of the most modern speculation about elementary particle physics that a putative role has been suggested for more than one 'family' of constituents. The muon was the first, and for a long time unappreciated, signal that there is a certain cyclic repetitiveness in the pattern of fundamental physics.

The other meson of the two-meson hypothesis fully deserved its name. The pion had strong interactions and did what Yukawa said it ought to do. As the postwar generation of accelerators came into operation the study of pions was greatly aided by their artificial production in the laboratory.

How can one sum up the feel of high-energy physics? Andrew Pickering, in his account of the evolution of the quark theory of matter, has described the style of the subject as being 'opportunism in context'.[16] The immediate postwar period, to which the greater part of this chapter has been devoted, certainly illustrates the opportunistic character of much of the research. Exploiting the bounty of cosmic rays, utilising wartime techniques for pure research purposes, daring to cross out infinite contributions and replace them with finite constants, all these have about them the air of being the acts of skilful and innovative people, bold enough to do what they can, and to considerable effect. Where Pickering and I disagree is in the assessment of the nature of the context in which this opportunist enterprise is being conducted. For him it is largely socially created. In his view the constraints of technical possibility and the tacitly agreed ways of thought adopted by the invisible college of the high-energy physics community conspire to *construct* (his word) the resulting picture of the nature of matter. For me the fundamental context is the way the physical world actually is, in its stubbornly idiosyncratic character. In my view, our picture of the nature of matter arises from the *discovery* (my word) of how, in fact, it behaves. We shall have to return to this issue after the story has unfolded. For the present I am content to suggest that the extraordinary accuracy of quantum electrodynamics has about it the air, not of an ingenious folk tale, but of a tightening grasp of an actual reality.

At the beginning of the postwar, pre-Rochester, period, the American Physical Society had convened a meeting in New York concerned with cosmic rays, accelerators and elementary particles. Its minutes recorded that, 'Disparate as these three subjects appear to be, the trend of physics is rapidly uniting them.'[17] The words proved to be truly prophetic. Remarkable progress was achieved in the period 1945–50 by their coming together.

Quantum electrodynamics was the most outstanding success. By the time of the third of the Shelter Island sequence of conferences its triumph was clear. Oppenheimer felt that thereby the original aim of these conferences had been achieved, and so that particular initiative, entirely American-based, came to an end. To the lasting benefit of physics, Bob Marshak was not content to let the matter rest there.

1   *Rochester 1*

In l950 Bob Marshak, an energetic young theorist, became the newly appointed Chairman of the Physics Department in the University of Rochester, in New York State. His department operated one of the new synchrocyclotrons, with enough energy available to produce pions. The department also contained a very active group studying cosmic rays. Thus Rochester University provided a most suitable vantage point from which to survey the progress of high-energy physics, and Marshak was determined to capitalise on this opportunity. On the 16th December 1950, a one-day meeting took place in his department. It was to prove to be the first of a continuing sequence of Rochester Conferences devoted to high-energy physics.

The Shelter Island series had been compact (about twenty-five participants each time), largely theoretical (with only two or three 'token' experimentalists present) and almost exclusively American. Marshak saw that the new era in particle physics, which the growth in sophisticated experimental facilities was bringing about, called for a different approach. He accordingly felt that, 'a new series of conferences should be inaugurated in which the experimentalists should be given 'equality' with the theorists.'[1] In view of the largely experimentally led character that the subject was to have during the greater part of the period we are surveying, this was a timely decision. The Rochester Conferences were also to be larger in their attendance and eventually (though this took some time to achieve satisfactorily) they were to be fully international in scope. The high-energy physics community is a sort of worldwide intellectual village and the Rochester Conferences were to become its town meetings. For a generation of high-energy physicists, their first invitation to attend was a gratifying signal that they were recognised on the international scene. (Towards the end of the period under review things changed somewhat. The Conferences had by then become large and unwieldy and in consequence the invitations particularly coveted tended to be to more specialised meetings of more manageable size, which concentrated on a specific set of problems.)

Not all theorists relished the change in character in their subject reflected in this new style of conference. Some of those of the older generation would have preferred to remain with the intimate, elite type of theoretical gathering as at Shelter Island. Paul Dirac has spoken of, 'what I think is a rather general principle in the development of theoretical physics; namely one should allow oneself to be led in the direction which the mathematics suggests.'[2] He, and people of his persuasion, were to remain rather detached from the high-energy physics scene during its long period of experimental dominance. There is a story, doubtless apocryphal, of Dirac giving a lecture in the 1970s in which he referred to the muon as a 'recently discovered particle'. Even some of the theoretical developments proved unpalatable. Dirac disliked the device of removing infinities by renormalisation, dismissing it as 'just a rule of thumb that gives results.'[3] I think that is unfairly derogatory, but it illustrates, if in rather an extreme form, the physicist's instinctive rejection of a purely instrumentalist account of his subject (all we need is a method of proven predictive power) in favour of enquiry with a realist intent (we want to understand what is going on).

The proceedings of later Rochester Conferences were produced and published (nearly always with astonishing rapidity) to provide valuable state-of-the-art surveys of the current position in particle physics. No such report, however, resulted from the first meeting. I have to rely on the brief resumé provided by 'Bram Pais in his book _Inward Bound_. Fifty people assembled for the occasion. Financial support had been provided by local industry, notably the Haloid Corporation, later the Xerox Corporation. There were three sessions, chaired respectively by Pais, Oppenheimer and Bethe, and devoted to accelerator results on the interactions of pions and nucleons, the physics of muons, and results from cosmic rays. Already, quantum electrodynamics was sufficiently well understood to be treated as routine, not to say old hat. Pais remarks that at Rochester 1, 'the subject of quantum electrodynamics did not even come up (although Feynman was there).'[4] There is an insistently instant character about high-energy physics. Once something is understood, let us hurry on to the next problem. I remember, when I was about to pay my first visit to the United States as a postdoctoral Fellow, an older English physicist said to me, 'They won't be interested in what you have done, only in what you are doing.' He was right. When one is young this rapid change of focus is exhilarating; as one gets older it becomes a little exhausting.

One topic which had begun to be of interest, and which would not prove to yield quickly to investigation, was the interactions of pions and nucleons. Although Yukawa had to some degree modelled his meson theory on the photon picture of electromagnetism, the triumphs of quantum

20

electrodynamics were not to be repeated. The reason was clear to see. Quantum electrodynamics' success stemmed from the fact that one could do reliable approximate calculations by expanding in powers of α. That worked because α was so small. The corresponding parameter for the pion–nucleon problem is not at all small; rather it is of the order of unity. This dashed the weapon of perturbation theory (the technical term for such approximate expansions) from the hand of the theoretical physicist. It seemed that there was precious little else available to put in its place. The problem of how to cope with strong interactions will be a continuing theme in our story.

One result which must have given satisfaction to the participants of Rochester 1 was the confirmation, achieved in the course of 1950, of the existence of the neutral pion (π^0). Such a partner to the charged pions (π^+ and π^-) was, of course, an essential ingredient in Kemmer's charge-independent scheme involving isospin. Only the charged pions had for sure been detected in cosmic rays. The π^0 was more elusive, not only because neutral particles are usually more difficult to detect than those which are electrically charged, but also because the neutral pion is a much shorter-lived particle than its charged brothers. The latter decay through their own version of the weak nuclear force. This gives them a lifetime of 2.6×10^{-8} seconds, which is very long by the standards of the subatomic world. A particle moving with a velocity comparable to the velocity of light can traverse metres in such a time and, in fact, charged pions travel much farther than that in the Earth's atmosphere since the 'clock' that tells them when to decay is moving with them, and according to special relativity such moving clocks run slow. The neutral pion, on the other hand, decays via the much stronger electromagnetic interaction into two photons

$$\pi^0 \rightarrow \gamma + \gamma,$$

in a notation which is transparent provided one knows that the third letter of the Greek alphabet always represents a photon (a gamma ray). This faster mode of decay gives the π^0 a lifetime of about 10^{-16} seconds. The existence of the neutral pion was first unequivocally demonstrated at the Berkeley synchrocylotron, through the coincidence of two appropriate gamma rays resulting from the decay. I might mention that even the π^0's lifetime is pretty long compared with the natural timescale of strong-interaction physics, which is of the order of 10^{-23} seconds. In consequence, particles like pions are conventionally listed as 'stable particles' in the tables which record particle properties.

Pais sees the discovery of the π^0 as a watershed in the changeover from a particle physics led by cosmic-ray experiments to a particle physics which accelerators came more and more to dominate. In 1950 'for the first time a new particle was discovered in an accelerator experiment.'[4] He obviously does not count J.J. Thomson's discovery of the electron using the 'accelerator' of his day, the cathode-ray tube.

2   *Rochester 2*

Just over a year after the first Rochester gathering, a second conference took place there on January 11–12th, 1952. It was supported by 'a group of Rochester industries' and attended by 41 locals and 52 'out of town guests'. Among the latter only Rudolf Peierls from Birmingham had an address foreign to the United States. He had come to Britain before the war as a refugee from the Nazis.[1] A theorist of distinction, he belonged to an omnicompetent generation, equally at home in high-energy physics, nuclear structure or the theory of the solid state. In the later years of our period these last interests were to dominate his thinking, but in the early Rochester years he was an active and significant figure in the international world of high-energy physics and an important link between it and the postwar generation of young British theoretical physicists. His institute at Birmingham attracted many talented people, including a number of postdoctorals who went to savour its rather down-to-earth atmosphere after the more rarefied and mathematical experience of doing a PhD at Cambridge.

The second Rochester Conference did not allow its proceedings to go unrecorded but the faded duplicated typescript of those proceedings is not now easy to find. I could not locate a copy in Britain and I had to go to CERN at Geneva to consult one. In his introduction to the document Marshak writes, 'In their account of the proceedings Mr A.M.L. Messiah and Dr H.P. Noyes [the scientific secretaries responsible] have attempted to combine haste, brevity and accuracy while at the same time retaining some of the flavour of the original discussions'.[2] They succeeded admirably. The accounts of all the early Rochester Conferences have great charm. They faithfully record the mixture of humour and seriousness so characteristic of the conversation of theoretical physicists when they are not forced into solemnity by too public an occasion. There is a cut and thrust to the discussion, with speculation and criticism interlacing each other. Anyone tempted to think of physics as an austere and desiccated activity would be disabused of that opinion by reading these very human documents. Later on, as the Conferences became very big, the atmosphere inevitably changed. The formality of procedure, necessary for coping with several hundred conferees, was reflected in a corresponding formality in the presentation of the proceedings, with the carefully prepared manuscripts of the review talks dominating their contents. Such discussion as these later volumes record has been carefully edited for publication.

Two very significant topics dominated Rochester 2. The new synchrocyclotrons were producing quantities of results about the interaction of pions with matter. The first day was given over to the questions this gave rise to. In the morning a rather confused picture was presented of the scattering of pions by various sorts of nuclei. The theoretical picture was bound to be cloudy since the analysis demanded an understanding of how nucleons are bound together to form nuclei as well as how pions interact with individual nucleons. Both high-energy physics and low-energy nuclear structure are tied together in the phenomena. One needs to know, for example, whether a proton and a neutron tend to stick together within a nucleus in the form of a deuteron subsystem, or whether they do not. At the end of a long consideration of such questions, Vicky Weisskopf delivered himself of the opinion that, 'he had not understood what either Oppenheimer or Bethe had said and did not think he was alone.'[3] There is always a temptation in difficult, half understood discussion to collude with what is going on for fear of making a fool of oneself by disclosing one's bafflement. People of directness and integrity, like Weisskopf, play an important role in being sufficiently confident to hint from time to time that the emperor might have left some of his clothes behind, or at least is wearing some of them back to front.

Everything is much easier if one can discard the obfuscating complexities of nuclear physics and concentrate on the interaction of a pion with a single nucleon. The simplest way to do this is to use the protons present in a container of liquid hydrogen. One needs the liquefaction in order to get a dense enough target to produce some actual events. Hydrogen gas is just too tenuous to be suitable. Of course, one would also like to know how pions interact with individual neutrons also. This could only be done in later experiments. There is no neutron equivalent of the container of liquid hydrogen. Instead, one must use a trick. A deuteron is formed by the very loose binding of a neutron to a proton. If one scatters off liquid deuterium it is possible to subtract out the known proton effects to leave the unknown neutron effects. The calculation is not trivial, however, and it needed an important theoretical development to make it possible. The proton and neutron get in the way of each other within the deuteron and Roy Glauber had to show how to allow for this shadowing correction when making the subtraction.

The session on pion—proton interactions was presided over by Enrico Fermi, perhaps the last great man able to straddle the experimental—theoretical

24

divide in high-energy physics. His department in the University of Chicago
was not only a centre for experimental investigation but was also a Mecca
for young American theorists of the highest talent. Nearly all the great names
in an outstanding generation of postwar American theoretical physicists spent
some of their formative years at Chicago. I remember George Steiner once
telling me, in the course of a conversation in a London taxi, that as a young
man at the University of Chicago in the late forties he had contemplated the
possibility of a career in theoretical physics. The cloud of talent around Fermi
scared him off. Physics' loss was literary studies' gain.

In contrast to the morning's obscurity, the afternoon session presented
a clear and interesting picture. The most striking result was that the ratio of
the cross-sections (that is to say, the ratio of the effective sizes of the 'obstacles'
producing the scattering of the pions) for the three processes occurring, namely

$$\pi^+ + p \rightarrow \pi^+ + p,$$
$$\pi^- + p \rightarrow \pi^0 + n, \qquad\qquad [1]$$
$$\pi^- + p \rightarrow \pi^- + p,$$

was 9:2:1. Keith Brueckner was at hand to offer a significant interpretation
of this fact. It proved to be the coming-of-age of isospin. At last people grasped
the fact that it was more than an ingenious device to implement charge
independence. Isospin was to become the basis of a new spectroscopy.

Physicists were familiar with the idea of adding together the angular
momenta of parts of a system in order to obtain states of specific total angular
momentum of the whole. Since angular momentum is conserved in interactions
with rotational symmetry, states which correspond to the same total angular
momentum will have the same properties. (Such states differ from each other
only by the value of the component of total angular momentum along some
direction, and if no direction is of preferred significance then neither is the
value of the corresponding angular-momentum component.) These states of
specific total angular momentum are made up by combining together states
corresponding to the angular momenta of the parts in certain fixed proportions.
These proportions (technically, they are called Clebsch–Gordon coefficients)
are fully determined by the underlying mathematics (technically, they derive
from the theory of representations of the rotation group). This angular-
momentum analysis was the basis of an old-style spectroscopy: the association
of states in families (or multiplets, as we say) with similar properties. For
example, the rules for adding angular momenta are such that if we add an
angular momentum of ½ to an angular momentum of 1 we get either a total
angular momentum of 3/2 or ½. (The units are in terms of the natural scale
provided by Planck's constant, $\hbar$.) This means that in an interaction involving these two angular momenta there are only two distinct possibilities for

25

what is going on, corresponding, respectively, to what is happening to total angular momentum 3/2 or ½.

Brueckner realised that isospin conservation followed from Kemmer's invariance under isospin 'rotations' and, moreover, since angular momentum and isospin are from the mathematical point of view identically structured (they are both rotations), the combination rules would be the same in both instances. Thus alongside the old-style spectroscopy of angular momentum could be set the new-style spectroscopy of isospin. The idea was immediately applicable to the interactions [1].

The proton has isospin ½, the pion has isospin 1. Therefore the total isospin of their combination is either 3/2 or ½. The significant fact that Brueckner had spotted was that the pure isospin 3/2 state gave exactly the ratio 9:2:1 observed in the experiments. Pure isospin ½, on the other hand, would have yielded the totally different ratio 0:1:2. Brueckner was able to conclude that in the region explored by the Chicago experiments the isospin state 3/2 was totally dominant in pion–nucleon scattering. (It was the only thing going on.) The value of isospin analysis was made clear, and from then onwards such considerations were to form a regular part of the discussion of strong-interaction processes.

Before leaving the subject we should note two important differences between angular momentum and isospin. The former not only occurs in the form of spin (an intrinsic angular momentum associated with each particle) but also in the form of orbital angular momentum (arising from the relative motion of the particles). There is no counterpart to this orbital component in the case of isospin. The second significant difference is that the conservation of isospin only holds in strong interactions. Electromagnetic or weak interactions do not possess the necessary 'rotational' invariance. For instance, the π^0 is a particle of isospin 1 which decays into two photons, which have no specific isospin. Therefore isospin is not conserved in electromagnetic interactions. On the other hand, all known interactions conserve angular momentum.

In October 1951, at an American Institute of Physics conference, Fermi had reviewed meson physics and had warned 'that we must be prepared for a long hard pull'[4] in seeking to understand it. Rochester 2 was an important step in what was indeed to prove a prolonged endeavour. In that afternoon discussion Oppenheimer had said that it was 'the first time there is any smell at all of phenomena which have been in the books for ten years.'[5] He was referring to calculations in so-called strong coupling theory, an idea now only of antiquarian interest, which had suggested that there might be excited (that is, higher mass) states of nucleons, just as there were excited, higher mass, states of nuclei. The first such state had been predicted to have

26

isospin 3/2, and angular momentum 3/2 also. It was tempting to suppose that this was what was influencing pion scattering. The excited state would be playing an unstable, intermediate role, being formed by the incident pion sticking to the proton and then the combination very rapidly disintegrating to give the outgoing pion and nucleon observed as the result of the scattering. This formation—disintegration process is the classic picture of what is called resonance scattering. It produces a large effect, manifested as a 'bump' in the cross-section. This bump centres on the mass of the unstable intermediate particle but it is spread out because the resonance particle's short lifetime does not allow it to have an exact value for its mass, due to Heisenberg's uncertainty principle linking energy and time. The shorter the lifetime, the wider the bump will have to be to satisfy Heisenberg. There was some discussion of this resonance interpretation at the Rochester session, but there was a good deal of resistance to it. Fermi was unenthusiastic. If the effect were of this origin the unstable particle would have to be very short lived to produce the necessary width of bump. Brueckner estimated its lifetime at 10^{-23} seconds. This is what we now think of as a typical strong-interaction time scale, but it was so much shorter than the time scales then familiar from work with resonances in nuclear physics that Gregory Breit, an acerbic expert in the latter subject, said that to call so broad an entity a resonance 'does not mean very much'.[6] More experimental work and theoretical acclimatisation were necessary before further progress could be made.

The second day of Rochester 2 was devoted to an entirely different, but at least equally significant, topic. This story began in Patrick Blackett's laboratory in Manchester in 1947. Two experimentalists working there, Clifford Butler and George Rochester, discovered two unusual events in the cloud chamber they were using for cosmic-ray investigations. One was a forked track, or V, which was naturally interpreted as the decay of an unseen neutral particle into a charged particle pair. The mass of this hypothetical neutral particle appeared to be large, of the order of a thousand electron masses or so. The other event was a track with a kink in it. That was reminiscent of the recent discovery of $\pi \rightarrow \mu$ decay, the kink being caused by the transition from one type of charged particle to another, the change of direction being due to an unseen neutral particle carrying off momentum. Again, however, the mass of the parent particle seemed to be considerably greater than that of the pion. These two events were signals of a new and entirely unexpected level in the structure of matter, but at the time they created little excitement. Pais says:

> In later years a discovery like Rochester and Butler's, however preliminary the data, would at once have caused excited corridor talk and would have

set the telephones ringing across continents. In 1948, I must have heard of the Manchester pictures but do not recall their causing a stir or immediate awareness of a new era being upon us in respect of the structure of matter.[7]

In a sense, that curious lack of excitement has continued. There can be few significant initial discoveries which have gained less honours for their discoverers.

Because of the form of the tracks, Rochester and Butler called these new entities 'V-particles'. The first thrill of discovery was followed by a long and depressing blank period. The particles were too massive to be found anywhere but in cosmic rays and for two whole years no further specimens obliged by appearing. Rochester has reminisced about how tantalising and embarrassing that desert period was for the Manchester group. Eventually, the pace of discovery quickened again. Cleverly triggered cloud chambers began to find more examples of V-decays and in 1949 the Bristol emulsion group reported the first case of a three-pronged decay of a particle labelled τ, of about 900 electron masses. These events were interpreted as being the decay of the τ into three pions. Here again the next confirming event was slow to appear. A year elapsed before two more τ's were found. At the first Rochester Conference Oppenheimer had suggested that there might be a discussion of the τ-mesons but he found no takers.[8] By the time Rochester 2 came around the scene had changed. A whole day was given over to the new particles.

The discussion took place under the odd title of 'Megalomorphs'. The chairman, Oppenheimer, reported that at an earlier meeting in Chicago, 'Fermi said he had become bored with the name "elementary particles". So the above name has been coined suggesting something with vast structure ... We hope to sharpen up the question of whether these objects are really astonishing'.[9] The rather preposterous name lingered on for a while but it never really caught on. A quantity of new data was available at Rochester 2, some of it coming from Caltech from 'the famous cloud chamber of Anderson that discovers all new particles'.[10] (It had previously notched up the positron and the muon.) The Bristol results had to be presented at these sessions by the theorist Peierls.

It was abundantly clear that the new particles were indeed there, but in how many varieties and with what properties were still the subjects of much confusion. One problem emerged with some clarity. It had a familiar ring to it. As detection techniques improved it became obvious that the new particles were fairly copiously produced. Yet they also lived for a fairly long time before decaying, since they were able to make observable tracks in cloud chambers and emulsion stacks. That was perplexing. The strong interactions

28

must obviously be responsible for the plentiful production and they might equally have been expected to induce a rapid decay. Was one seeing a rerun of the earlier Yukawa meson puzzle? There is always a temptation for physicists, like generals, to refight the battles of the last war. It seldom works. On this occasion the strong−weak paradox was to find a resolution quite different from that of a second two-meson hypothesis.

Instead of invoking a heavy and copiously produced parent (quasi-π) which then gave birth to a weakly interacting child (quasi-μ) the solution lay with the invention of a new selection rule. Selection rules are constraints imposed on what can happen by the need to conserve some particular quantity. The conservation of electric charge provides a simple example. A neutral particle cannot decay into an odd number of charged particles, since there must always be as many negative fragments as positive fragments in order that the total charge comes out as zero. The quantity conserved in a selection rule is conventionally called a quantum number (sometimes, to emphasise the valuable property of its conservation, it is called a good quantum number). No known quantum number seemed to help with the V-particle problem, so Pais was bold enough to invent one for that purpose. With the slightly proprietary encouragement of Oppenheimer he explained it to the Rochester Conference. It was 'mass number' (N) and Pais suggested that it took the value 0 for ordinary particles, like pions and nucleons, and the value 1 for the exotic new V-particles. A rule was imposed which required that in a strong-interaction process the evenness or oddness (whichever it was) of the sum of the mass numbers of the participating particles should not change in the course of the interaction. That meant that if one started off with ordinary particles (in, say, a nucleon−nucleon collision) then even if there was enough energy available to create a V-particle the rule would not allow it unless there was also enough energy to create a second V to go with it. Otherwise, the evenness of the initial state would have been changed into oddness. The heart of Pais's idea was that it enforced associated production, that is to say that V-particles could only be produced from ordinary matter in pairs. When it came to decay, however, each particle was on its own. The weak interactions responsible for these decays could be supposed not to respect the N-rule. By divorcing production from decay Pais removed the perplexity about why the former was copious and the latter was dilatory. They were simply completely different processes. [A little reflection showed that Pais's rule was better understood as involving a multiplicative quantum number. Such quantum numbers take the values ± 1 only and they are combined, not by adding, but by being multiplied together. If one writes $n = (-1)^N$, then $n = 1$ for ordinary particles and $n = -1$ for Vs. The conservation of evenness/oddness is the requirement that the combined value of n is the same before and after.]

Oppenheimer turns to talk to Bethe, while Marshak in the front row looks on.

The other side of Rochester 2. Fermi is at the extreme right.

Pais's particular way of enforcing associated production did not in the end prove to be quite the way that nature had chosen to operate, as we subsequently shall see. Yet the idea of associated production was a capital one, as was the general intuition that the new particles were going to require new selection rules. Pais also wanted to use his new quantum number to produce a difference between the electron (to which he assigned $N = 0$) and the muon (to which he assigned $N = 1$). That was not such a successful notion, although the underlying hunch that somehow the structure of hadrons and the structure of leptons were related was to surface again very much later in a very different form. It is not clear that Pais deserves very much credit on that score (it is rather like hitting a target after a succession of lucky ricochets) but he certainly deserves great credit for the idea of associated production and the recourse to novel selection rules. He spoke prophetically when he said that he 'would like to look at the schematization as the unfolding of an ordering in which one talks about families of elementary particles rather than the elementary particles themselves'.[11] He received magisterial endorsement from Oppenheimer when the latter said that, 'This is a way of beginning to take these things seriously and beginning to codify them'.[12]

The plot of elementary particle physics was beginning to thicken. Its early innocent simplicity was getting lost as the experimentalists revealed the intimations of a 'vast structure'. I have written elsewhere:

> Progress in understanding the structure of matter is made in a spiral fashion. First some constituents are identified. Then further investigation reveals that a few more are required to do justice to the complexity of the phenomena. After a while, as their number increases, the situation becomes distinctly unaesthetic. A plethora is threatening. Just when all seems lost, a structure is perceived which in due course is interpreted as being due to the existence of a small number of constituents at a yet more fundamental level.[13]

Although those attending Rochester 2 did not know it, they were at the start of just such a spiral of discovery. The principal subjects of their deliberations, isospin and new selection rules, were to be absolutely crucial to the elucidation of what was going on in this new phase of the exploration of the structure of matter. Yet the physicists attending the conference did not at all feel like stout Cortes surveying with amaze a vast new ocean of knowledge. Annexed to the proceedings is a versified 'Summary' by Arthur Roberts. Its 73 lines are too long to reproduce in full, but these extracts will convey its tone. No doubt they reflect the spirit in which the conference dispersed:

> Once upon a time the world was less complex
> With fewer nervous wrecks, with lots more time for sex.

Electrons and protons were all that we could afford.
They were good enough for good old Rutherford.
The positron, the neutron and the horrible neutrino,
The beginning is easy to recite for us,
The ending is nowhere in sight for us,
And though the answer will sometime be nearer,
Things will get worse, before they get clearer.

We had pi-mesons and mu-mesons
And some thought it too few mesons,
Went out and discovered new mesons.
Some people don't know when to stop.

We have weak coupling, strong coupling,
Wrong as we-knew-all-along coupling,
Each month our troubles are quadrupling,
Some people don't know where to begin.[14]

3 *Rochester 3*

1952 is the only year to have had held in it two Rochester Conferences, the second in the January and the third on 18—20th December. By the end of the year the importance of these conferences was already becoming clear and the foreword to the Proceedings of Rochester 3 says, 'In view of the great number of topics discussed at this Third Annual Conference and the interest shown in earlier Proceedings, it was decided to make this year's Proceedings generally available at a nominal cost.'[1] Clearly, many took advantage of this opportunity, since from now on copies of the proceedings are a standard part of any serious physics library. Initially, they were produced in a simple style by the University of Rochester but they were distributed by commercial publishers (Interscience). The same foreword says that, ' Thanks are due to the Gray Audiograph Company and the IBM Company for supplying the recording equipment and typewriter respectively.'[2] From such small acorns do great oaks grow! Another development was that this was the first conference to attract government funding, via the National Science Foundation.

The first session of Rochester 3 was concerned with nuclear forces. Its chairman was Eugene Wigner. An expatriate Hungarian, powerful mathematical physicist, eventually to be awarded a Nobel Prize for his work on the application of group theory to relativistic physics, and brother-in-law of Dirac (who is said once to have introduced his wife with the words, 'Have you met Wigner's sister?'), he was famous in the physics world for his exquisite, not to say excessive, politeness. The quintessential Wigner story concerns his returning a totally unsatisfactory used car to the swindling dealer with the words, 'Go to hell, please.' I remember, as a young man, being introduced to Wigner and being completely unnerved by the wholly undeserved greeting, 'Such a great pleasure to meet you, Dr Polkinghorne.'

Two topics dominated the discussion, in this first session, of the force between two nucleons. One was the application of isospin. The fundamental origin and nature of this internal symmetry remained mysterious and Wigner said that its significance was a subject likely to recur again and again. He

34

allowed himself an uncharacteristically sharp remark in relation to isospin selection rules, saying that 'there has never been as much theoretical thinking done on a subject, the experimental foundation of which was as inadequate as this one.'[3] The remark was greeted with loud laughter, but the last laugh was to be with those who were bold enough to exploit these new ideas.

The second topic was that of what potential was to be associated with the nuclear force. Over the following years enormous amounts of time and effort were to be expended on this question. From a fundamental theoretical point of view it was a misconceived exercise. Of course, a potential could be a useful phenomenological device, a calculational tool for use in discussing nuclear structure or low-energy nucleon−nucleon scattering experiments, but it was no more than that. Since the goal of physics is not mere prediction but is actual understanding, the question of a nuclear potential was a distinctly secondary issue. The very idea of a potential belonged to an outmoded thought-world. It represented an interaction which would be mediated instantly, in contrast to relativity's requirement of an influence propagating with a velocity less than or equal to that of light − a picture that was literally realised in Yukawa's concept that it was the emission and absorption of mesons which brought about the nuclear force. Not that, however, the latter picture proved at all tractable when it came to attempting actual detailed calculations.

The candidate potential that received most attention in late 1952 was due to a Frenchman, Maurice Lévy, who sought to extract it from a more fundamental base by overingenious argument. Oppenheimer commented on Lévy's proposal that 'it is not completely clear from his papers, it is not completely clear to him and not completely clear to anyone.'[4] In his summary, Oppenheimer said, 'starting with a not unreasonable theoretical program and making only a finite number of mistakes, Lévy has obtained a better overall charge symmetric description [of the data] over a wide range of energies than people who have been treating the problem empirically.'[5] He was gracious enough to say that he thought this was not without interest.

It is characteristic of the theoretical sessions at these early Rochester Conferences that they contain a good deal of reasonably good-natured ribbing of one speaker by another. The experimental sessions have about them a greater air of earnest seriousness.

Much detailed progress was made in 1952 with the investigation of pion−nucleon scattering. The strategy was to analyse the experimental data in terms of phase shifts.[6] These latter correspond to the natural way of parametrising the data, a way that automatically respects unitarity — that is, it fulfils the basic requirement that in a scattering experiment something must emerge again. More formally, phase shifts ensure that no probability is lost. If there are a variety of possible outcomes, the sum of the probabilities associated with them must add up to one — i.e. something must happen. There is a phase shift (expressed as an angle) associated with each state of total angular momentum and total isospin. Provided that the scattering does not permit the creation of extra particles in the final state (as will be the case if it is below the energy threshold for such particle creation and which will only occur to a small extent at energies not too far above that threshold) the phase shifts are always real. At these lower energies only a few angular-momentum states will be important, since slow-moving particles cannot possess large angular momentum in the finitely-sized region in which the scattering takes place. All in all, phase shifts provide a very convenient form in which to analyse the data without making detailed assumptions about what is happening. They may be used to identify the presence of resonances. A phase shift δ gives a contribution to the cross-section which is proportional to $\sin^2 \delta$. This is, of course, a maximum (gives a bump) when $\delta = 90°$. A phase shift passing through $90°$ therefore generates a resonance peak.

Fermi's group set to work to analyse the $\pi-p$ data in this way. They found that the phase shift corresponding to total angular momentum (J) 3/2 and total isospin (I) 3/2 passed through $90°$ at about 180 Mev energy. They had discovered the celebrated (3,3) resonance.[7] Study of the way in which the scattering varied with the scattering angle could then disentangle which orbital angular momentum (l) went to make up the total angular momentum of 3/2. It could have been either $l = 1$ or 2, since either possibility added to the proton's spin of ½ would give $J = 3/2$. It turned out, in fact, to be $l = 1$, or a p-state as we say. The phase-shift analysis was greatly assisted by using the early generation of computers. It was proudly proclaimed at Rochester 3 that, 'with the use of an electronic computer the phase shifts can be computed in five minutes'[8] — another acorn from which a great oak was to grow! Fermi told of how the results of the analysis, performed on a Los Alamos computer, were 'handed him on a small piece of paper on which was written, in cryptic fashion as is proper for something that comes from Los Alamos, certain numbers which had to be decoded.'[9]

According to the picture of resonance scattering, it proceeds by the formation and decay of a highly unstable intermediate state. That meant that the Chicago group in discovering the (3,3) resonance had, in fact, discovered a new particle, quite as worthy of the name as a proton or a neutron, even if it was so very short lived (10^{-23} seconds) in the enjoyment of its particlehood. It took a long time, however, for physicists to think of it that way. Pais comments that, 'It took quite a few years before physicists became comfortable with the idea that there is no real difference between a resonance and an unstable particle.'[10]. There was reluctance to admit the possible onset of a plethora of particles.

For a while there was also another difficulty. There was a competing set of phase shifts on offer which fitted the data available with equal efficiency, but which did not show up a resonance. It was a classic case of the underdetermination of theory by experiment. Frank Yang was able to show that there was always an alternative set of phase shifts which would fit the kind of data then available. They might look less attractive theoretically (they lacked the simplicity of a single dominant state) but they did the empirical job as well as Fermi's. Only later on, when further discriminatory experimental information came to hand, did it become clear that Fermi's intuition had not led him astray. The (3,3) was really there.

The theorists were already addressing pion−nucleon scattering on the basis that this would prove to be the case. One of them was a young man called Geoffrey Chew. When he talked at Rochester 3 we are told that he 'began this discussion with what he characterised as a simple-minded theoretical attempt to understand the problem on the basis of Yukawa's fundamental idea.'[11] The notion was to concentrate on the basic Yukawa process of the emission or absorption of a single pion by a nucleon. Since the latter was seven times heavier than the pion, it was not a bad approximation to treat the nucleon as being so massive that effectively it was fixed, 'nailed down' and nonrecoiling.

The idea is mildly technical but worth exploring. The basic process

$$n \rightarrow n + \pi \tag{1}$$

has total angular momentum ½ in the initial state (it can only come from the fixed nucleon's spin); angular momentum conservation then implies that it must take the same value in the final state. The pion is spinless and so the angular momentum of the final state comes only from the nucleon's spin together with the orbital angular momentum of the emitted pion. It follows that the latter can only be either $l = 0$ or $l = 1$ (for if one had, say, $l = 2$, then this, added to the nucleon spin, would give either $J = 3/2$ or $J = 5/2$, both of which are too big). Further progress can be made by considering parity. This is a multiplicative quantum number (see p. 31) associated with behaviour under reflection. At the time, it was believed to be conserved in all interactions; it is still believed to be conserved in the strong interactions that Chew was discussing. Each particle contributes its own factor (its intrinsic parity). For the nucleon this is $+1$; for the pion, which is what is called a pseudoscalar particle, this factor is -1. There is also a factor of $(-1)^l$ corresponding to orbital angular momentum. Applying parity to expression [1] and collecting together initial and final factors, which must be equal to each other, yield the equation

$$+1 = (+1).(-1).(-1)^l, \tag{2}$$

from which one impeccably deduces that $l = 1$, rather than $l = 0$. In other words, the pion emitted in a Yukawa process by a static (fixed) nucleon is a p-wave pion.

At this point a light goes on in the theorist's mind, for Fermi's analysis points to p-wave pions as dominating in the region of the (3,3) resonance. Chew went on to use a potential-type approximation (the language of the day) to argue that his model led to a resonance in the isospin 3/2 state. We are told that he 'had agreed to make the following rather glib statements only with the understanding that Dyson and Bethe would not contradict him at this session, but would take up these points in the technical theoretical session.[12]' When that session came, more elaborate theoretical schemes were indeed discussed, with another young theorist, Francis Low, very much to the fore. The intellectually up-market idea was called the Tamm—Dancoff approximation, an attempt to work selectively in terms of the numbers of particles present in a state. It proved a pretty intractable technique and prone to be plagued with infinities. All in all, the big theoretical guns did not succeed better in hitting the target than Chew's attempt with bow and arrow. Everyone was groping — in the right place, as it turned out, but not yet with the right point view to guide them.

The other great subject discussed at Rochester 3 was the continuing saga of the V-particles. Some sort of order was appearing as the number of recorded instances grew into three figures. Attention concentrated on a clump of mesons with masses around 900 times that of the electron. The only way of labelling them was in terms of their decays. There were the taus, decaying into three pions; the kappas, which decayed into a muon and two unidentified neutrals; the chis, which decayed into a charged pion and one unidentified neutral. Results came from both cloud chambers and emulsions. The Frenchman, Louis Leprince-Ringuet commented on the latter. He was permitted to do so in French provided it was 'so clear that all will know it'. He paid a generous tribute to Powell's group at Bristol.

> En Europe, il y a pour les émulsions, Bristol, le grand soleil, et puis un tout petit nombre de petits satellites dont la dimension, même en faisant la somme, reste très inferieure de celle de Bristol.[13]

In later conferences, Leprince-Ringuet was constrained to attempt English, a performance which someone (Oppenheimer, I think) characterised as 'statistical bilingualism'. He was certainly right about the Bristol group's dominance at this time. Of 12 European τ events, seven came from there, three from London and one each from Padua and Rome. Examples of V-particles were still so rare that individual events were lovingly discussed in great detail, with due attention being paid to the analysis of grain densities in the films and the like.[14] One is reminded of the *Beagle* era in zoological exploration, in which expeditions brought back new specimens for taxonomic investigation. There is an old-world charm about this stage in the development of high-energy physics.

Much of the Conference's attention was centred on the basic question of what was going on. Feynman asked about evidence for associated production and was told that there was little, though Bruno Rossi conceded that this might be due to the presence of decay modes which were wholly neutral and so escaped attention.[15] One important phenomenon was exhibited at Rochester 3. There were now enough data to produce plots showing lifetimes associated with the various different decay modes. It was astonishing to find that they all overlapped. It was the first clear sign that these decays might have something in common. Fermi commented, 'It is a striking plot and probably has a meaning too, I would say.' Oppenheimer then expressed the hope that, 'our great grandchildren when they attend the 2038 [*sic*] Conference in Rochester will take it for granted that they understand these things.'[16] The plot did indeed have a meaning and one did not have to wait until the twenty-first century before discovering it.

4 *Rochester 4*

The foreword to the Proceedings of the Conference held in Rochester on 25–27th January 1954, states that the Conference had 'the good fortune to follow a year in which a substantial amount of new experimental material on high-energy physics became available.'[1]

One area in which that was certainly true was nucleon–nucleon scattering. Not only were higher energies available for investigation but also more detail could be obtained about what was happening. It became possible to study spin-dependent effects. How two nucleons interact can be expected to depend in part on the mutual relationship of their spins. The simplest experiments had simply averaged over the possible spin orientations. Now people began to do double-scattering experiments, in which one interaction is followed by a second with a further target. This second interaction can be used as an analyser of the amount of polarisation (spin difference) produced by the first interaction. The polarisation effects were found to be large. This made it necessary for the 'potential', still the object of many theorists' thoughts, to become much more complicated (technically, it was required to have a large non-central component). At the same time, the limitations of the potential approach were beginning to be recognised more clearly. Breit acknowledged that at the highest energies then being investigated in detail for p–p scattering (400 Mev), 'We are working on a sounder basis if we confine ourselves to phase shifts.'[2]

The discussion of meson scattering was trying to get more deeply theoretical in its treatment. It was, however, proving exceptionally difficult to make successful contact between the phenomena and basic theory. There was, for the latter, a quantum-field theory to hand, describing the interaction of pseudoscalar pions with nucleons in a charge-independent way. This theory was known to be renormalisable, and so it possessed all the virtues that current wisdom decreed. It expressed in proper relativistic quantum-mechanical form Yukawa's idea of meson–nucleon interaction. The problem was how to extract empirically meaningful consequences from it. We have already noted the difficulty presented by the size of the coupling constant, which did not permit a convincing treatment using perturbation theory. There was grave difficulty even in deciding precisely what value should be assigned to that coupling constant. How could you measure it except by reference to experiment? But if you could make no experimental predictions from your theory, how was such a comparison to be made? It was an impasse. Without a chicken, how could one get an egg? Without an egg, how could one get a chicken?

It was instructive to consider how the corresponding problem was solved for quantum electrodynamics. It depended on the (J.J.) Thomson limit. This was an example of what came to be called a low-energy theorem. One could show that for photons of almost vanishing energy their cross-section for scattering by an electron was given exactly by an expression proportional to the square of α and with a known coefficient. Obviously, this put one in the business of measuring the fine-structure constant. A session of Rochester 4 was devoted to seeing if one could do something like this for pions as well as for photons. The upshot was not encouraging. A natural counterpart to the Thomson limit was not identified, and there was the additional problem that, because of its non-zero rest mass, one never has a pion with an energy less than mc^2 anyway. The debate was confusing and confused. We are told that it generated 'much heat but no light'.[3]

Rochester 4 was the last Conference attended by Fermi. By Rochester 5 he was dead from stomach cancer at the early age of 53, an enormous loss to physics. Towards the end of Rochester 4 he said that he had broken precedent with himself by attending a theoretical session because there was a point that 'shall I say gave me hope'. Mr Goldberger [a young theorist of whom more will be heard] was making a gallant attempt at killing the pseudoscalar theory. Fermi went on to say that such an assassination would be 'a great boon to physics, second only to a definite proof that the theory had something to do with experiment'.[4] That perfectly sums up the prevailing attitude to fundamental meson theory in 1954; one could neither quite live with it nor without it.

During 1953 the Cosmotron, the somewhat pretentiously named first large working synchrotron, had come into operation at Brookhaven. In consequence, Rochester 4 was able to receive results on pion−nucleon scattering up to 1.5 Bev[5] in energy. It was reported that at around 1 Bev there was a second broad peak, interpretable as a resonance, but this time the charge ratios indicated that it was occurring in the $I = \frac{1}{2}$, rather than $I = 3/2$, state. The cloud no bigger than a man's hand, which the (3,3) had represented, was beginning to grow. Soon it would result in a deluge of short-lived hadrons.

Meanwhile, some dust was settling in the area of the V-particles.

Two classes had been identified: those heavier than nucleons, which were called hyperons and which would turn out to be spin-½ particles, and those intermediate between pions and nucleons, which were given the generic name of K-particles (later changed to K-mesons when it was recognised that they were of integer spin). There was still a good deal of uncertainty about how many different hyperons and Ks there were, and about their detailed properties. 'Megalomorphs' had made its exit from the terminology, and the summary table of the proceedings related to these new particles was headed instead 'Curious Particles',[6] in acknowledgement of their perplexing and unexpected properties. An important aid to the analysis of one type of K-decay was unveiled at Rochester 4 by Dick Dalitz. A down-to-earth Australian, he had disliked his student days in Cambridge but had begun to flourish when he transferred to the more matter-of-fact setting of Peierls's Birmingham. Dalitz was thinking about τ decays. Because there are three particles in the final state, they are quite complicated to analyse. Dalitz had hit upon the perfect way of displaying the results, now universally called a Dalitz plot. If this two-dimensional plot were to be uniformly populated, the orbital angular momenta involved in the final state would all be zero. The state would then have to derive its parity solely from the intrinsic parities of the three pions, that is it would have to be negative. Any clustering in the plot would suggest the presence of higher angular momenta, with a consequent modification of the parity discussion. Dalitz's original plot contained eleven unambiguous τ decays. The preliminary conclusion was that, 'The small number of events doesn't allow strong conclusions, but the data does look rather uniform.'[7] It was recognised that this $J(P) = 0(-)$ attribution for the angular momentum (parity) of the τ would be different from the corresponding attribution for another K-particle, the θ°. The latter was defined by its decay into π^+ and π^-. If it had spin zero, the two pions would have to be in a state of orbital angular momentum zero, giving the $\theta^\circ J(P) = 0(+)$. Thereby was to hang a very considerable tale.

Rochester 4 was the first of these Conferences to be written up in *Physics Today*, the chatty house-journal of the American Institute of Physics, whose articles are intended to be accessible to all physicists, whatever their speciality. The account was compiled by Pierre Noyes, a rather pragmatic sort of theorist, and it is garnished with rather attractive line drawings of some of the principal participants by Rona Finkelstein.[8] Noyes summed up the state of play in meson scattering by saying:

> the pseudoscalar meson theory is still with us in the sense that it has always been, namely as a theory incapable of yielding rigorous results in our present state of mathematical ignorance, and readily yielding nonsense when clumsily handled, but still showing contact with reality at enough points to keep us from junking it altogether. Although the general atmosphere of theorists was

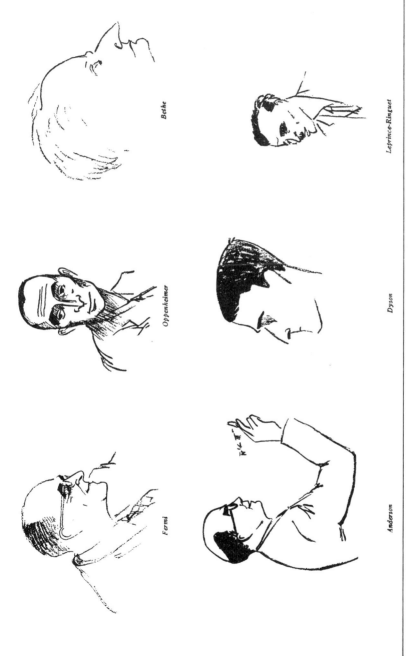

Bethe

Leprince-Ringuet

Oppenheimer

Dyson

Fermi

Anderson

Rona Finkelstein's charming sketches of some of the great men who attended Rochester 4.

gloomier than at last year's conference, there is still a considerable body
of opinion that the difficulties inherent in getting reasonable answers from
the theory are not insuperable.[9]

That was a judicious verdict. Less happy was Noyes's pronouncement on
the curious particles. 'Theoretical attempts to reconcile ease of production
of these particles with their subsequent lifetimes are still at the speculative
stage.'[10] People were cautious because cosmic rays afforded so little clear
evidence for associated production. Confirmation of that idea would require
something more systematic than haphazard findings in balloon-flown emulsions
or cloud chambers on mountain tops. Work at machines like the Cosmotron
would soon make up the deficiency. Meanwhile, it is ironic that 1953 had
seen the publication of what proved to be the definitive theoretical solution
of the problem.

It came from an up-and-coming young American theorist, Murray
Gell-Mann. He is one of the heroes of our story, highly creative in his
theoretical thinking but always with a weather-eye open to what the
experimentalists might be saying. Murray once told me how he acquired his
healthy respect for experiment. It was when he was a graduate student at MIT.
He went to a seminar in which an expert on the theory of nuclear structure
was giving an account of a detailed calculation he had done concerning the
shell-model prediction for the spin-parity of some particular nuclear level.
Let us say that he triumphantly concluded it must be $2(+)$. At the end, a
little man at the back of the room put up his hand. 'I work in the basement,'
he said, 'and I've just measured the spin-parity of that level. It's $1(-)$.'
Collapse of theorist, and the lighting of a lamp in Murray Gell-Mann's mind
that has burnt there with considerable illuminating power ever since; always
talk to the experimentalists in the basement.

In August 1953 Gell-Mann published a paper in which he made his
first great permanent contribution to physics. We are still at a time when 'East
is East and West is West and never the twain shall meet', so it is no surprise
that a similar idea was conceived and published a little later by a Japanese
theorist, Kazuhiko Nishijima. They had both found the right way in which
to mix isospin and a new quantum number in order to explain what was going
on with the curious particles.

The two nucleons form a doublet with isospin ½, and the three pions
form a triplet with isospin 1. One can write a formula for the electric charge
(Q) corresponding to a given isospin component I_3 in a conventionally
chosen '3' direction

$$Q = I_3 + \tfrac{1}{2}B, \qquad\qquad\qquad [1]$$

where B is the baryon number, which counts the number of strongly interacting

44

spin-½ particles; that is, B is zero for the mesons and is 1 for the nucleons (and subsequently for the hyperons also). (Until very recently it was believed that B is absolutely conserved. The contemporary speculation about very rare violations of B-conservation is of no relevance to the present discussion.) The formula [1] works. For example, π^- has $I_3 = -1$ and $B = 0$, giving $Q = -1$; the proton has $I_3 = ½$ and $B = 1$, giving $Q = 1$; and so on.

The Gell-Mann−Nishijima scheme placed the curious particles in isospin multiplets but the then-known hyperons (unlike the nucleons) were to be assigned integral isospin, and the K-mesons (unlike the pions) were to be assigned isospin ½. In order to avoid the incorrect consequence of fractional charges, the relation [1] had to be modified in these cases. That is, where the new quantum number comes in. With its invention, Gell-Mann inaugurated the long story of facetious terminology in particle physics. He chose the name 'strangeness', and soon the curious particles became the strange particles, which they have remained ever since. The later history of particle physics is littered with such fossilised jokes.

The key idea was that strangeness (S) modified eqn [1] to read:

$$Q = I_3 + ½(B + S). \tag{2}$$

For ordinary particles of zero strangeness eqn [2], of course, reduces to eqn [1]. The new hyperons were assigned strangeness -1. There was a neutral isospin-zero particle, the Λ°, and a triplet corresponding to isospin 1, the Σ^+, Σ° and Σ^-. Positively charged K-mesons were part of an isospin doublet with isospin ½ and strangeness $+1$; negatively charged K-mesons were part of another doublet, with isospin ½ but strangeness -1. Strangeness was assumed to be conserved in strong and electromagnetic interactions (the latter was necessary to prevent the fast decay $\Lambda \rightarrow n + \gamma$), but the weak interactions did not conserve strangeness so that, for example, the Λ could eventually decay into a proton and a π^-, as indeed it did.

Associated production was an immediate consequence of the scheme. Hyperons, like Λ or one of the Σs, had to be produced in company with a positive strangeness K. A negative strangeness K-meson, which we will denote by $\overline{K}$ (since it is the antiparticle of a positive-strangeness K) would also have to be produced in association with a K, a process which has a higher energy threshold than that for hyperon production. All these consequences of the scheme were in due course found to be the case.

In assigning a neutral companion Σ° to the known Σ^+ and Σ^- hyperons, Gell-Mann and Nishijima were hypothesising the existence of a new particle. One could readily understand why it had not then been seen, for the strangeness rules permitted its fast electromagnetic decay into the lighter Λ plus a photon. The existence of the Σ° was only confirmed in 1956. By then I was a postdoc at Caltech. I remember Gell-Mann and Feynman looking excitedly at the bubble-chamber photograph from an accelerator experiment in which the first Σ^0 was discovered. The Λ had manifested itself by its decay into $p + \pi^-$; the photon showed up by converting, that is by interacting with the matter in the chamber to produce a visible electron–positron pair. When one added up the energy-momenta of the decay products, the sum corresponded to just the right mass to be a Σ. At first, Gell-Mann and Feynman were puzzled that the photon had been able to convert in the chamber. Then they noticed the details of the apparatus which revealed that the bubble chamber had been filled with the heavy liquid freon (also used in refrigerators). Its density made such electromagnetic interactions much more likely than they would be in a chamber filled with a light liquid like liquid hydrogen.

About the time of the proposal of the strangeness scheme, evidence began to come in from cosmic-ray experiments of another kind of new particle. It exhibited a cascade decay, that is to say it first changed into a Λ plus a pion and then the Λ underwent further decay into a nucleon and a pion. Assigned the symbol Ξ, these cascade particles were readily accomodated in the scheme as a doublet of isospin ½ and double strangeness, $S = -2$. That gave the Ξ^- (which had been seen) and required it to have a partner Ξ^0 (not then yet discovered). The cascade form of decay was explained by the supposition that in weak interactions strangeness could only be lost one unit at a time. A doubly strange particle like the Ξ had first to change into a singly strange particle like Λ, before finally shedding all strangeness in the subsequent Λ decay.

One further consequence of the strangeness scheme is worth noting. The neutral K^0 has an antiparticle $\overline{K}^0$ which is different from itself (since it has the opposite strangeness) unlike, say, the π^0 which is its own antiparticle. We shall find that eventually some very interesting consequences were recognised as following from this differentiation.

5 *Rochester 5*

The Fifth Rochester Conference, held from the 31st January to the 2nd February 1955, was the first to succeed in attracting a significant international attendance. Of the 136 participants, more than twenty came from Europe, of whom seven were from Britain. One of the latter was Abdus Salam, whose graduate student I then was at Cambridge. The overseas contingent might have been smaller but for prompt action by Bob Marshak. He recalled that 'we experienced for the first time the evil consequences of the McCarthy era.'[1] It was also the time of Oppenheimer's trouble with the Atomic Energy Commission and his security clearance had been revoked a few months earlier. Marshak says, 'his presence at the conference introduced political overtones and animated discussion outside the conference halls which had never existed before.'[2] In such an atmosphere it was not surprising if some of the more left-wing European physicists who had been invited fell foul of the McCarren–Walter Immigration Act, which had been designed to exclude communist influences from the United States. The authorities insisted at first on treating a short visit as if it were a bid for permanent residence. Marshak went to Washington, accompanied by a Rochester Congressman, and succeeded in persuading the State Department to sponsor the necessary waivers from the US Attorney-General.

I remember very well Salam's return from the Conference. He was excited by what he had heard and he brought back a clear message. Something of real promise was at last beginning to happen in the long-arid area of the theory of strong interactions. It centred on the work of Chew and Low. They were a strikingly contrasting pair: Geoff Chew, ingenuous, enthusiastic, a real positive thinker; Francis Low, subtle, sophisticated, saturnine. They were colleagues at Urbana, Illinois, but it was almost inevitable that eventually Geoff should gravitate to the West Coast and Francis to the East. Meanwhile, from the lowbrow direction of Chew's static model and from the highbrow direction of Low's manipulation of field theory, their work had converged on the recognition that the crude treatment of the static model, in achieving its success in affording an understanding of the (3,3) resonance, could be seen as the low-energy limit of a local field theory. Phenomenological success and intellectual respectability were at last beginning to embrace each other. Exactly what was going on was still not yet fully appreciated. Ideas were floating around which eventually would come together to provide the basis for a powerful assault on the fortress of strong-interaction dynamics. Chew said of Low's equation that it 'almost fulfils, for the Yukawa theory, the Heisenberg idea that all one ever needs to talk about are elements of the scattering matrix.'[3] Marvin Ruderman stressed the importance of the so-called optical theorem (which relates the scattering cross-section to the imaginary part of the forward-scattering amplitude). Abraham Klein referred to the idea of crossing (a relation between different processes, first identified by Gell-Mann and Goldberger). Peierls talked about the use of the complex plane to associate resonances with poles off the real energy axis. All these notions are important ingredients of what came to be the S-matrix approach to strong interactions. They will be more fully discussed when the ideas attain a more articulated form at later Conferences.

In the meantime, the experimentalists were beavering away. One of their projects was to try to measure really low-energy pion—nucleon scattering, a difficult task which produced results which were described in mathematical joke-language as 'integrable only in the sense of Lebesgue.'[4] In plainer words, they were pretty erratic, the reverse of something expected to vary smoothly with energy. Higher up the energy scale, processes involving production of extra pions in the final state were also proving pretty perplexing. Yang said, 'so far attempts at understanding have consisted in "cataloguing" the observations.'[5] This 'natural-history' stage of the accumulation of 'for instances' is often a necessary preliminary in the exploration of a new and complicated regime. In nucleon—nucleon scattering the endlessly ramifying construction of potentials continued its unfruitful way.

Two important advances in available facilities had taken place in the preceding year. Another synchrotron, the Bevatron at Berkeley, had come on the air with sufficient energy to achieve fairly frequent production of strange particles. This opened up the opportunity for the systematic exploration and confirmation of their properties. The future in this respect lay with the machines, as C.F. Powell had recognised as early as 1953. He had concluded his remarks at a cosmic-ray conference with the words, 'Gentlemen we have been invaded . . . the accelerators are here.'[6]

The other advance was the increasingly successful development of bubble chambers now that people had found out how to make them 'almost pathologically clean',[7] a necessity in order to avoid boiling at the boundaries. Luis Alvarez, one of the kings of bubble-chamber physics, presented an account of a liquid-hydrogen chamber he had constructed at Berkeley. Like that same laboratory's first cyclotron it was just four inches in diameter. Alvarez, a man given to thinking big, would soon be planning a 72″ chamber. A great industry was in the making. Handling the analysis of the vast numbers of photographs would require semi-automatic scanning and measuring, together with computerised storage of data. Pais says,

> I well recall the initial reservations of quite a few members of the profession to Alvarez' managerial approach which, they felt, would remove physicists yet another step from experimentation done in the old ways. They were right and so was Alvarez. It came to be recognised that either one accepts the new style as indispensable, or particle physics will languish. How else could the Berkeley group alone have measured 1.5 million events in 1968?[8]

Associated production of strange particles was still not well-established experimentally. When the Indian physicist M.G.K. Menon came to survey the situation he could write down only nine assured instances of its happening. A participant offered from the floor another example to add to the list. Pais in his theoretical review would not go beyond saying that he felt 'the experimental situation is more encouraging than when he first suggested the idea'.[9] He went on to refer to 'an attempt by Gell-Mann of assigning quantum numbers to particles which seems to show promise'.[10] Caution was the watchword of the day. It warms my heart, a little later, to read that, 'At this point Pais interrupted his talk to make a plea not to use words like strange or peculiar (referring to new particles). For a theoretician, a proton ought to be as peculiar as a hyperon. This is only terminology, but one might bear it in mind.'[11] Alas this was a vain hope! Pais concluded his remarks with the question, 'Is there some comprehensive principle of symmetry, a beginning of a dynamics, behind all this?'[12] There was indeed, but it would be some time before that became clear.

Meanwhile, the K-meson mystery was deepening. All measurements of masses and lifetimes seemed to be converging on identical values, whichever decay mode was investigated. It was extraordinarily difficult to see how such exact coincidences could come about if there was more than one type of particle involved. Already, in 1953, at a cosmic-ray conference, Bruno Rossi had commented on the emerging identity of the τ and θ masses, 'This hardly looks like an accident.'[13] At Rochester 5, Dalitz had 53 τ decays recorded on his plot. His conclusion was, 'If the spin of the τ meson is less than 6 it cannot decay into two π mesons.'[14] In other words, unless one assumed a preposterously high value for the spin, in terms of current wisdom the τ and θ just could not be the same, masses and lifetimes notwithstanding. Dear, oh dear!

In the course of his talk, Pais had mentioned that the weak decays of strange particles were characterised by the I_3 component of the isospin changing by $\pm\frac{1}{2}$, a rule which is summarised by

$$|\Delta I_3| = \tfrac{1}{2}. \tag{1}$$

Meditation on eqn [2] of Chapter 4, taking into account the conservation of B, shows that this is just a rephrasing of the requirement that strangeness only changes by one unit at a time. Gell-Mann immediately proposed from the floor that this should be generalised to the stronger requirement that, to a considerable degree of accuracy, the total isospin should change only by $\frac{1}{2}$

$$|\Delta I| = \tfrac{1}{2}. \tag{2}$$

Gell-Mann and Pais later wrote a pioneering paper on this rule. The restriction [2] is more powerful than is [1] and it helps to explain some otherwise puzzling facts. The argument is a little involved, but is worth pursuing. For example, the θ^+ decay

$$\theta^+ \longrightarrow \pi^+ + \pi^o \tag{3}$$

takes place much more slowly than the apparently analogous neutral decay

$$\theta^o \longrightarrow \pi^+ + \pi^-. \tag{4}$$

The predominance of the condition [2] offers an immediate explanation, for it allows [4] but forbids [3]. (The reason is that Bose statistics requires the two πs in [3] and [4] to be in a symmetric state. It is easily seen that this in turn requires their isospin state to be either $I = 0$ or $I = 2$. The former is possible for [4] and agrees with [2] for the decay of the $I = \frac{1}{2}\,\theta$, but [2] has to be $I = 2$ (since it is a charged state) which [2] forbids.) The rule [2] must be dynamical in origin, as its approximate, if predominant, nature indicates that it is not a precise symmetry rule (as [1] is). In the years intervening since Rochester 5, enormous progress has been made in many areas of particle physics, but no convincing explanation has yet been offered of why [2] works so well. Murray Gell-Mann wrote recently, 'It never occurred to me that more than thirty years later the reasons for that approximate $|\Delta I| = \frac{1}{2}$ rule would still be the subject of research.'[15]

An innovation in procedure was introduced at Rochester 5. High-energy physics was growing ever more extensive and more complex and so, in order to accommodate the amount of new material available, it was necessary to split the Conference according to specialisms and to hold some sessions in parallel. It was therefore found desirable to have a summary talk at the end to draw the threads together. This was to become a regular feature of Rochester Conferences until the continually increasing scope of the subject

An impressive front row — Freeman Dyson with three Nobel Prize winners (Steinberger, Feynman, Bethe).

made the attempt by any one person to give a global view too difficult a task to contemplate. On this first occasion the summary was given by Dyson. Commenting on the varied and often conflicting insights proferred by contemporary theory, he said, 'to summarise a theoretical session is a contradiction in terms.'[16] It was clear, nevertheless, what the keynote had been. It was the same as that message that Salam had brought back to Cambridge: 'The grand success has been the Chew model which considers the nucleon to be a non-relativistic object.'[17]

In 1954, two important events had happened which were off-stage as far as Rochester was concerned. On the 29th September 1954, sufficient ratifications were received from the Western European participating nations (including the United Kingdom) to bring into being CERN, a collaborative international centre for nuclear research. The subject was becoming big business and big business needs big money. The USA and the USSR might be able to go it alone; the more modestly sized countries of Western Europe needed to get together if they were to continue their long history of successful involvement in the investigation of the structure of matter. With hindsight, one might regret that the chosen location for the venture was Geneva — an expensive milieu with an ethos of diplomatic-style living — but without CERN the countries of Madame Curie, Wilhelm Roentgen, Enrico Fermi, Niels Bohr, J.J. Thomson (and the adopted home of Ernest Rutherford) would have been condemned to impotent spectatorship during one of the most exciting and fruitful periods in the history of physics.

The second event was the publication by Frank Yang and Robert Mills of the first modern (non-abelian) gauge theory. The gauge idea itself went back to Hermann Weyl in the 1920s. It was his way of thinking about electromagnetic interactions. It is a notorious fact about quantum theory that the wavefunction, $\psi(x)$, is not uniquely, physically determined. Nothing in the physical consequences of the theory is changed if we multiply $\psi(x)$ by a factor of modulus unity

$$\psi(x) \longrightarrow \exp(i\lambda) \cdot \psi(x). \qquad [5]$$

Here λ is just a real number and so it is, of course, the same whatever point x is considered. A transformation of this type is called global; it is the same everywhere. Weyl considered the possibility of replacing expression [5] by a local transformation; that is to say, one in which the phase transformation is different at different spacetime points x. The exponent λ must then become a function of x

$$\psi(x) \longrightarrow \exp(i\lambda(x)) \cdot \psi(x). \qquad [6]$$

Now it is no longer true that nothing changes. In the Langrangian (the expression which specifies the physical nature of the system) there are differential operators which will act on $\lambda(x)$ and produce factors of $d\lambda/dx$ all over the place. Only if these factors can be soaked up in some way will local gauge invariance (no change under eqn [6]) hold. Weyl saw that electromagnetism enabled one to achieve this. The electromagnetic fields are derived from a vector potential, $A_\mu(x)$ (the suffix μ denotes that A transforms like a four-vector under the transformations of special relativity). It was well known that the physical fields do not determine the corresponding A_μ uniquely. There is an ambiguity and the fact that physical consequences are independent of how it is resolved is now called gauge invariance. This ambiguity could be exploited to soak up those unwanted terms in $d\lambda/dx$, by incorporating them into a permissible redefinition of A_μ.

To the mathematically uninitiated this argument may seem a little complicated. The essential point is this: Global invariance (everywhere changing in the same way) can be turned into local invariance (different changes at different points) provided there is a vector field around (in this case the vector potential of the electromagnetic field) to remove the terms which otherwise would be left over.

Yang and Mills had the idea of trying the same trick for isospin. Its 'rotations' were originally global. If charge independence turns a proton into a neutron it does so everywhere. Could this be generalised into a local symmetry, with a proton becoming a neutron here but not there? The answer was that it could, provided that once again there were some vector fields around to mop up the mess. There would have to be three such fields: one for each isospin component. Because 'rotations' do not commute with each other (one can easily see that the order in which they are performed matters) the mathematics became rather more complicated than in Weyl's case. The electromagnetic field does not itself carry electric charge, but the new isotopic vector fields had to carry their own isospin and so interact with each other. In the technical language of mathematics, Yang and Mills had invented a non-abelian gauge theory. Exactly the same idea occurred to a Cambridge graduate student, one of my contemporaries, a clever but slightly fey young man, Ron Shaw. He did not publish this work, other than in his PhD dissertation, because he feared that the theory had a fatal flaw. One of the consequences of gauge invariance for electromagnetism is that it guarantees the massless character of the photon. That is excellent, but Shaw felt that the same might also be true for these new vector fields. That would be disastrous, for one certainly did not see massless, strongly interacting particles around the place. Only much later was a way found out of this dilemma.

It was not obvious at the time, but the invention of non-abelian gauge theory provided the prime tool for future developments in fundamental theory.

Such gauge theories are usually called Yang—Mills theories. Only the most learned and fair-minded add —Shaw. Looking back, I am surprised that Shaw's supervisor, Salam, did not manage to persuade him to publish. Salam is one of those bold thinkers, uninclined to let little formal difficuties get in the way of what might prove to be a good idea. (He once offered me the cheerful, if somewhat dubious, advice, 'Publish all your ideas — people will only remember those that turn out well.') On the other hand, Ron Shaw can be pretty stubborn when he has made his mind up.

6 *Rochester 6*

Nearly two hundred people attended the Sixth Rochester Conference on April 3rd—6th 1956. Sixteen scientific secretaries were required to cope with the organisation and recording of the sessions. The Proceedings were prepared at the Institute for Advanced Study, Princeton, in collaboration with Princeton University and Brookhaven National Laboratory. It was all becoming a big enterprise and that reflects itself in the more formal style that was adopted for the proceedings. A corresponding refashioning of the actual sessions of the conference was slower in coming. People were loath to give up the right of the individual to present his own work so that for a while crowded and prolonged sessions became the order of the day. At the start of the discussion at Rochester 6 devoted to heavy mesons and hyperons, the chairman, Clifford Butler, 'purposely refrained from writing down the list of 15 or 20 contributions to the session to avoid frightening the audience.'[1]

In the year which had elapsed since the last Rochester meeting, the antiproton had been discovered at Berkeley. The energy attainable by the Bevatron had deliberately been designed to make this possible. There had been rumours that the occasional antiproton expected to turn up in cosmic rays had not obliged. Some speculations had begun in the relentlessly fertile minds of theorists about whether the Dirac partnership of matter and antimatter was as universal as had been supposed. No real progress had been made along these lines, which was just as well since the new accelerator soon provided unambiguous evidence that the antiproton was there, as expected, even if the experimentalists had to break down the wall of the Bevatron building and extend their detecting apparatus into the parking lot in order in order to confirm that this was so. The result — for which Chamberlain and Segré were eventually awarded a Nobel Prize and which later led to an unedifying squabble in the physics community about who suggested what when — was reported at an American Physical Society meeting in Los Angeles in late 1955. I was there and I remember the atmosphere as being one of sober satisfaction, rather than the excitement that a truly unexpected result would have produced. By the time Rochester 6 came round, four months later, antiprotons had become routine. High-energy physics is like looking-glass land; you have to move really fast to stay ahead. The cosmic-ray physicists were at hand to explain that they really had seen antiprotons after all, including what Rossi called a 'prehistoric' example from way back in 1954.[2] That was the trouble with cosmic-ray physics. Everything was seen (including quite a lot of things which weren't actually there), but not everything was immediately recognised. With its uncontrollable circumstance and low statistics, cosmic-ray studies approximated to what some philosophers of science seem mistakenly to have considered to be the typical character of physical investigation: the imposition of a grid of expectation upon a flexible set of data. In fact, accelerator physics, with its clear signal of a negative particle of protonic mass, is very much more the paradigm case of how science proceeds. But let's be fair to the cosmic-ray people — they did identify the V-particles before machine-based physics could begin to investigate them.

Much progress had been made with those strange particles by Rochester 6. The Σ° and Ξ° were now certified members of the club, the latter being described as 'now well known even though there are only about 10 known cases'.[3] No question remained about the fact of associated production, but as to its interpretation, Oppenheimer would not permit himself more than the somewhat grudging comment, 'This problem has reached a temporary kind of solution in the theory of strangeness.'[4] Frank Yang was warmer. He acknowledged that the last year had seen 'the firm establishment of the "strangeness" quantum number'.[5] Among its various successes was the exclusion of the process

$$n + n \rightarrow \Lambda^\circ + \Lambda^\circ, \qquad\qquad [1]$$

which the multiplicative quantum number of Pais would have allowed, but which was not observed. Yang referred to this aspect of affairs as the 'convergent part of the problem'. There was also a 'divergent part of the problem',[6] where matters were far less satisfactory. That concerned, of course, the deepening perplexities about the τ/θ puzzle (see p. 42). It seemed as though one was being painted into a corner by the continued convergence of the masses and lifetimes associated with the different decays. Yang remarked, 'Of course, if they are all different decay modes of the same particle the puzzlement would vanish. ... However ... Dalitz's argument strongly suggests that it is not likely that [τ and θ] are the same particle.'[7] The problem was parity. By now Dalitz had accumulated 600 points on his plot, which exhibited 'a remarkably uniform distribution'.[8] Something funny was going on. Oppenheimer uttered a characteristically gnomic epigram: 'The τ meson will have either domestic or foreign complications. It will not be simple on both fronts.'[9] In the discussion, Feynman brought up a question that he attributed to an experimentalist, Marty Bloch, presumably too shy in theoretical company to ask it for himself, 'Could it be that the θ and τ are different parity states of the same particle which has no definite parity, i.e. that parity is not conserved.' Yang gave a cautious reply. He 'stated that Lee and he looked into this matter without arriving at any definite conclusions'. Gell-Mann expressed himself as being in favour of an open mind. As the discussion continued, 'the Chairman [Oppenheimer] thought the moment had come to close our minds. . . .'[10] Before the session ended he loosed off another sybiline sentence. 'Perhaps some oscillation between learning from the past and being surprised by the future is the only way to mediate the battle.'[11]

Thereafter, things moved rapidly. I recall Feynman and Gell-Mann returning to Caltech from Rochester, both very excited over the question of parity non-conservation. Feynman, in particular, insisted that we did not really know what happened in this respect in weak interactions. Curiously, he did not go on to do anything about it. Pais records that, 'on the train back from Rochester to New York, Professor Yang and the writer bet Professor Wheeler one dollar that the theta- and tau-mesons were distinct particles'.[12] Yang did not long remain of that opinion. He and T.D. Lee quickly set to work to do a thorough analysis of what was known about the parity question. Soon Yang would be able to pay his bet from his share of the proceeds of a Nobel Prize.

Important progress had been made theoretically in tackling the physics of strong interactions. People had been able to identify the general idea of which Chew—Low theory, so strikingly successful at the previous conference, was a particular limiting example. The answer proved to lie in the complex plane of mathematics. In physics we are directly concerned, of course, with values of the energy which are positive, real numbers. It had long been known that in certain circumstances the mathematical entities appearing in physical theories could be extended in a natural way to complex values of the 'energy' (that is to say, the 'energies' could take unphysical values, even including factors of i, the square root of -1, and still give 'sense') and that such extensions were very well behaved. In more technical language, the physical functions were boundary values of analytic functions of a complex variable. That is a very powerful mathematical property for the functions to possess and it leads to expressions in which the function can be given in terms of the integral of its imaginary part over the real axis. Such expressions are called dispersion relations. From the 1920s they had been familiar to electrical engineers and the like, who were concerned with the responses of electromagnetic systems. For these systems the arcane mathematical property of analyticity could be shown to be equivalent to the physical property of causality (nothing comes out until something has come in; the bell does not ring before the button has been pressed) and vice versa.

The systematic application of these ideas to high-energy physics began in 1954 with a seminal paper by Gell-Mann, Goldberger and Thirring. They looked at the interaction of photons with nucleons and identified the input necessary for deriving a dispersion relation for the process. This input was a 'microscopic causality condition', expressed in quantum field theory as the technical requirement that the commutator of fields at spacelike separation should vanish. Although this is not so intuitively direct a condition as the macroscopic causality condition of the engineers described above — I cannot explain it in ordinary words — it is a natural way of expressing in the formalism the causal independence of spacelike separated events, just as relativity requires. GGT (abbreviations are the order of the day with multiple authorship) also realised that the discussion of the scattering of massive particles, like mesons, might be susceptible to a similar treatment. Soon this was in full swing.

The first dispersion relations were confined to forward scattering, that is to say, to the special case in which no momentum is transferred between the interacting particles. Owing to the optical theorem (p. 48) the imaginary part then appearing in the integrals is related to the physically measurable cross-section, at least for positive energies. The integrals also involved negative ranges of the energy. Here the idea of crossing (p. 48) came to one's aid. The notion can be summarised in somewhat cavalier fashion by saying that a negative-energy incoming particle looks the same as a positive-energy outgoing antiparticle, so that the negative-energy expression for the process

$$1 + 2 \rightarrow 3 + 4$$

can be replaced by the positive-energy expression for the 'crossed' process

$$1 + \bar{3} \rightarrow \bar{2} + 4,$$

where 2 and 3 have been interchanged, or crossed over, becoming antiparticles in the process. Gell-Mann and Goldberger had been the first to show that this crossing property holds in a relativistic quantum theory. As a consequence, forward-dispersion relations involved integrals over physical cross-sections, together with certain (single-particle) pole terms which proved to afford just the natural way of defining coupling constants (force strengths). (A pole corresponds to the particularly simple mathematical expression $1/x$.) At Rochester 6, Murph Goldberger was able to summarise the position achieved in these words.

> The point I would like to emphasise about these dispersion type equations is that they provide what may be regarded as exact predictions of field theory,

the testing of which will require much more accurate experiments than have heretofore been carried out. This is an unusual and gratifying position for the theoreticians to be in after all these years. I'm sure every red blooded experimentalist will want to rush back to his laboratory in an effort to produce data to contradict the equations.[13]

There is no mistaking the confident tone of the theorists after so many years of frustration in the face of strong interactions. By the time of the Conference many separate groups had seen that an extension from the forward direction to the discussion of processes with non-zero momentum transfer should also be attempted. However, a price had to be paid for that. The dispersion integrals no longer involved only measurable quantities. There was an unphysical region at low-energy, which inexorably appeared in the integrals but which was inaccessible experimentally. (That was due to the fact that in order to have some momentum to transfer, the particles necessarily had physically to have the energy resulting from that momentum. In the laboratory you could not get as low in energy as the mathematics required.) One was faced with the difficult problem of how to extrapolate into this never-never land of unphysical values.

Not only was there this difficulty at low-energy but there was also a problem at high-energy, which equally affected forward or non-forward relations. The integrals that resulted had to be convergent (that is, well defined) to make sense. Whether this was so or not depended on how the imaginary parts behaved when the energy became very large. The right way to think about this took some time to find. In the meantime people simply guessed.

The dispersion-relation programme not only afforded the chance for theory to make contact with experiment. For the physics of strong interactions it also offered the possibility of eventually replacing field theory with a more directly empirical formulation of relativistic quantum mechanics. On this view the dispersion relations were not thought of as 'exact predictions of field theory' but rather as potential substitutes for it. Heisenberg had emphasised that the interpretation of experiments did not call for the elaborate machinery of a local quantum field theory, purporting to describe what happened at every point of space and point of time. The regions in which elementary particles interacted with each other were not open to our inspection. All one actually determined experimentally were the relative probabilities for the different outcomes that could result from a projectile particle's incidence upon a target particle in specified circumstances. The rules of quantum mechanics enabled one to calculate these probabilities if one knew the elements of the S(cattering)-matrix, a set of complex numbers (technically, probability amplitudes), one of which was associated with each pairing of initial and final possibilities. The S-matrix elements were the quantities appearing in dispersion relations and the constraints which the latter imposed promised to give a much-needed sharpness to the Heisenberg programme of working in terms of S-matrix elements alone. Without the definiteness furnished by these analyticity requirements, Heisenberg's ideas looked rather vague. Now it seemed that they might prove an adequate basis for an articulate theory. Later on, this S-matrix programme became particularly identified with theorists working at Berkeley, but in the first dawn of dispersion theory it attracted much wider attention. Gell-Mann spoke in these terms at Rochester 6, whilst back home at Caltech he encouraged postdocs (like myself) to work on detailed aspects of the programme. His reminiscences express his perspective on the matter. He recently wrote:

> This was the dispersion theory program that I outlined at Rochester in 1956, pointing out that it was a way of formulating field theory on the mass shell (that is, in terms of physical quantities). At Rochester I showed how one could in fact obtain the whole set of scattering amplitudes on the shell [S-matrix elements] iteratively starting with the particle poles, but admitted I didn't know what the boundary conditions [at infinite energy] would have to be for a particular field theory. I casually mentioned Heisenberg's S-matrix program, almost as a joke, but later on after we had convinced Geoffrey Chew of the value of the program, he renamed it 'S-matrix theory'. As you know Landau (around 1959) and Chew (around 1961) insisted that it was somehow distinct from field theory.[14]

Meanwhile, the conventional field theorists continued labouring away. A toy field theory, not realistic but possibly instructive, had been invented by T.D. Lee. The Lee model suggested the ominous possibility of 'ghosts' in quantum field theory, that is to say, the occurrence of states carrying negative probability, a disaster which would destroy any possibility of a consistent interpretation of the theory. Was this just an artefact of the model or an indication of real trouble? Landau's group in Leningrad thought they had arguments to demonstrate the latter. At Rochester 6 these questions were discussed by the sharp-tongued Swedish physicist, Gunnar Källén. He said of the claim to have discovered ghosts in realistic field theory that, as in other spooky stories, 'though people who tell sincerely believe them, they are not necessarily true'.[15] Unfortunately, Landau was too independently minded to have been let out of the Soviet Union to defend his ideas at Rochester. Those compatriots of his who were present (who probably included a few bulky individuals whose names were not well known in the physics community) declined to risk speaking on his behalf.

On the experimental side the accelerator physicists were in undisguised triumphant mood. Robert Leighton from Caltech speeded his humbler colleagues on their way: 'next year those people studying strange particles using cosmic rays had better hold a rump session of the Rochester Conference somewhere else.'[16]

Two topics were on the agenda at Rochester 6 which would rumble on for many years. One was the study of hyperfragments (nuclei in which a nucleon was replaced by a hyperon). This unique combination of the difficulties of nuclear-structure physics with those of strange particle physics was to prove attractive to a dedicated subgroup of theorists.

Right at the end of the Conference, Robert Hofstadter from Stanford presented results of measuring the form factors of the proton. In its electromagnetic interaction the proton does not behave as a pure point particle and the form factors measure the way in which its charge and magnetic moment appear to be 'smeared out' over a region of space. Hofstadter was eventually awarded a Nobel Prize for the delicate investigations his group at Stanford carried out into these properties over a period of many years. At Rochester 6 he felt neglected and he concluded his report with the plaintive remark that the absence of comments made the group feel they had been 'sort of working in a vacuum'.[17] A good deal of interesting physics was learnt from these experiments, but ultimately the perception of their status changed. When one thought of the proton as elementary, then its electromagnetic form factors assumed a character of greater significance than they enjoy now that the proton is recognised as a composite system of less absolute fundamentality.

Before leaving Rochester 6 one should note that there was still a refusal to think of the (3,3) resonance as a *bona fide* particle. Weisskopf solemnly expressed the desire that no doubly charged particles should exist and concurred with Oppenheimer when the latter added the wish that no strongly interacting particle should have a spin exceeding ½.[18] Both these hopes were already contradicted by the (3,3).

7   Rochester 7

Rochester 7 was the first Conference that I attended, as one of the quite large delegation of British physicists. In those far-off days we travelled by transatlantic liner, our universities being unable to finance what was then the costlier option of air travel. Although the Conference dates were just the 15–19th April 1957, we were all away for the greater part of the month.

The central topic of the Conference was the non-conservation of parity in weak decays, which had been discovered in the year intervening since the last Rochester meeting. If interactions conserve parity then physics seen in a mirror is indistinguishable from physics observed directly. An immediate consequence is that there can be no preferred handedness in nature, since left and right interchange under reflection and the preference would change correspondingly. Conversely, the discovery of such a handedness is an unambiguous signal that parity is not conserved. Such signals had been observed in weak interactions. The first case involved the decay of nuclei of cobalt-60, with the emission of an electron and an antineutrino. The cobalt was surrounded by a coil carrying a current and the experiment was conducted at very low temperature to avoid the issue being confused by the effects of thermal fluctuations. The magnetic field due to the coil aligns the spin of the cobalt nuclei and this correlates their decay behaviour with the direction in which the current is circulating. One can define an 'up' direction with respect to the coil by specifying it to correspond to the way a right-hand screw has to point if it is to rotate in the same direction as the current. Correspondingly, a left-handed screw defines a 'down' direction. Mrs C.S. Wu found that more electrons were emitted 'downwards' than 'upwards'. Because of the handedness involved in defining these directions, it follows immediately that parity is not conserved in this weak decay. Soon more examples of parity non-conservation followed. These involved decays of pions and muons. The leptons emitted were found to be linearly polarised, that is to say their spins were aligned with their directions of motion. A moment's reflection convinces one that such an alignment corresponds to a preferred direction of twist about the direction of motion, which again is only possible if the decays are not even-handed. (Otherwise the left- and right-handed twists would balance each other out to give no net polarisation.) All these effects were observed to be large and unambiguous.

By the time Rochester 7 assembled these results were well known. All the interactions in which parity non-conservation had by then been found involved neutrinos. Even before these experimental findings, Salam and Landau had both realised that there was a particularly neat way of involving these massless particles in parity violation. Neutrinos have spin ½. Let us resolve that spin along the neutrino's direction of motion. The result is either + ½ (spin and momentum aligned in a right-handed sense) or − ½ (spin and momentum aligned in a left-handed sense). Of course one can also discuss linear-polarisation states of massive spin-½ particles like electrons, but relativistic kinematics implies in their case that the linear polarisation cannot be perfect, there is always some mixing of left and right states which cannot be kept wholly separate from each other. Not so for neutrinos. Their right- and left-hand polarisation states are quite independent of each other. Therefore one can have one without the other. In other words, a two-component neutrino theory is possible, with neutrinos purely left-handed (one component) and antineutrinos purely right-handed (the other component). Such thorough-going handedness clearly produces maximal parity violation for neutrinos, in a fashion that was later found to be consistent with the experiments.

No one supposed, however, that the whole blame for parity violation could be laid at the door of the neutrino alone. After all, the whole issue had arisen over the τ/θ puzzle where no neutrinos were involved. It was expected that this problem was now to be resolved in the way that Feynman had tentatively raised at Rochester 6. The coincident masses and lifetimes could be attributed to there being a unique particle involved, able to decay in either a τ or a θ mode because parity conservation was no longer there to forbid its doing so. Yet the two-component idea seemed attractive enough to encourage attempts to preserve it in some modified form even for the neutrinoless weak interactions. By Rochester 7 people were playing around with this idea but had not yet found the right way in which to implement it. Much difficulty arose from the fact that there were still a lot of wrong results around relating to the detailed character of nuclear beta decay. The experiments were tricky and were often to prove to have been erroneous. I remember, much later on, hearing T.D. Lee say that a funny sort of statistics seemed to hold in weak-interaction physics, where a claimed three standard-deviation effect (that is, one which should be very reliable) had a fifty per cent chance of being right!

Once conservation laws began to crumble, it was natural to ask where to stop. There were two other discrete transformations which were similar to parity (*P*) in the sense that they involved total transposition from one state to another, rather than the sort of gradual change that is associated with a continuous transformation like a rotation, which can proceed by a series of small steps. One such transformation was charge conjugation (*C*) which replaces particles by antiparticles, and vice versa. The other was time reversal (*T*) which involved something like the transposition of past and future. Of course, that particular interchange was not quite a feasible proposition, but an equivalent transformation was to reverse all velocities, since the reversed system would then develop forwards into the future in the same way that the original system would have developed backwards into the past. In the early fifties, mainly through the work of Pauli and a young German called Lüders, it was realised that all theories had to be invariant under the combined effect of all three transformations applied simultaneously. This was the celebrated *CPT* theorem, a deep and universal consequence of relativistic quantum mechanics, just as inescapable as the relation between spin and statistics. It followed immediately that you couldn't have *P* violation without violating at least one of the others in a compensatory way. It was soon realised that charge-conjugation invariance was being broken in weak interactions. Time-reversal invariance was much more difficult to check but it was seen as an attractive option that it should be conserved. Then violations would be confined to *C* and *P* alone in such a way that their product *CP* was conserved (since the *CPT* theorem implied this was effectively the same as *T*). One of Landau's chief motives for suggesting a two-component neutrino theory was that it elegantly incorporated this *CP* property. (It treated on an equal footing the left-handed neutrino and its *CP* partner the right-handed antineutrino.) Landau was not alone in finding it a metaphysically soothing thought that then one could not tell the difference between antiparticles seen in a mirror and particles observed directly. The desire to cling to some kind of reflection invariance was extraordinarily strong.

At the same time people were busy explaining that they personally had never thought that parity was anything special. Someone who was certainly above trading on the wisdom of hindsight and who truly held such a view, was Paul Dirac. When he was asked what he thought about parity violation he simply replied that parity was not in his book, the justly celebrated *Principles of Quantum Mechanics*. Apparently, he never believed reflection symmetry to be a fundamental property of nature. Gell-Mann was of the same opinion. He tells us that when he was a graduate student at MIT in 1949, in response to a work assignment, 'I turned in a statement that conservation of parity ... is an empirical law',[1] that is to say, something that may or may not prove to be the case. On the side of the experimentalists reactions were rather different. The experiments had been feasible for many years but no one had bothered to try them. Hard-luck stories circulated about people who had intended to do so but were, for one reason or another, deflected from their purpose.

A session of Rochester 7 was devoted to parity non-conservation. Yang was its chairman and Lee its principal speaker. Together they had been the prime movers in the affair, first establishing by careful analysis that parity conservation was an open question in weak interactions and then pointing out the critical tests which would settle the matter. Soon they were to be rewarded with a joint Nobel Prize. Oddly enough, Feynman (despite his enthusiasm) and Dalitz (the originator of the triggering τ/θ perplexity) had not used their theoretical powers to pursue the question to an answer. Lee's talk, which he prefaced with the assurance that 'the particular points of view I shall adopt are fully shared by Professor Yang',[2] speculated about possible T-violation. He was willing to entertain the idea that if it (and so CP) did not work out here on Earth, perhaps there was a cosmic variability of kinds of matter which would restore a global balance. There would still be a sort of reflection invariance provided one looked in a big enough mirror! Lee and Yang continued to work together in fruitful partnership for a number of years. I remember, during a visit to the Institute for Advanced Study at Princeton, hearing them chatting away in Chinese, with the occasional occidentally recognisable word like 'nucleon' or 'cross-section' popping up in their discourse. Later, to the sadness of many of their admirers, they split up. Eventually I was to hear T.D. Lee give a talk on the history of weak-interaction physics which was remarkable for the fact that it did not mention Frank Yang from start to finish. Neither was able to treat the other with generosity.

The strong-interaction theorists continued their exploration of the complex plane. Scattering amplitudes might be well behaved there but they could not be totally so without degenerating into triviality. An analytic function without any singularities (points of bad behaviour) can only be a boring constant. Since that would not do to describe the complexities of physics, it became necessary to identify what singularities were actually present. The first to be noted were called normal thresholds, points at which sufficient energy became available to permit the possibility of creating a further extra particle in the final state. The 'hiccup' represented by this new option produced the kind of singularity known to the mathematicians as a branch point. Normal thresholds occurred at real values of the energy and they enjoyed an immediate physical interpretation. It would have been very satisfactory if they and the single-particle poles had been the only singularities one had to reckon with. Schwinger presented a paper at Rochester 7 in which, in his rather high-flown style, he essentially made that claim for a particular set of amplitudes. In the Proceedings we are told concerning the aftermath of that particular talk that part of the discussion was lost.[3] I suspect that was a diplomatic move. I recall that as Schwinger's ringing tones died away, Källén rose to his feet. He said he didn't know very much about the problem but he knew enough to be able to say that the previous speaker was totally wrong. An instant chill descended on the meeting at this stinging rebuff delivered to a great physicist. Källén was right, all the same. The singularity structure of scattering amplitudes was to prove to be very rich and subtle, beyond naive expectation.

That was a topic I was later to spend several years working on. On this occasion I was to make my own first contribution to the conference.[4] Goldberger was the chairman and he had a blackboard listing the many speakers, together with the times allotted to them. Most were imbued with the importance of what they had to say, which did not allow them to take too seriously paltry restrictions on length of presentation. Murph Goldberger's genial nature did not lend itself to draconian chairmanship. As the session wore on, the board got smudgier and smudgier as figures were revised. Those junior speakers, like myself, placed at the end of the list found that, despite the lengthening of the session beyond its scheduled end, they were being squeezed into smaller and smaller slots. Frankly, it was all pretty chaotic. The session devoted to pion reactions had adopted a better procedure, with an exceptionally clear and well-organised talk by the Liverpool experimentalist Jim Cassels, summarising a good deal of material. The future lay with this method of approach.

Cassels popped up again in another session to give some experimental results about a rare pion-decay mode. He did so in the midst of a lot of very highbrow theoretical discussion. He described himself as 'the one plumber among ten millionaires', saying, 'Let me apologise right away for introducing an experimental peasant note into this very refined discussion.'[5]

Ideas were in the air at Rochester 7 which were later to prove significant beyond their impact at the time. In a session devoted to the structure of the nucleon, Yoichiro Nambu (from Japan but settled in the United States) suggested that sense could be made of some of Hofstadter's measurements if there were to be a very short-lived neutral meson. He was proposing the particle we now call the ω. It was a bold stroke, for which Nambu said he 'would take full responsibility for all the consequences'.[6] He was taken to task by the session chairman, Peierls, who ironically commented, 'You seem to know so much about this particle.'[7] Nambu is a quiet man, often hard to follow, who has made a number of important suggestions which did not command immediate recognition.

Another important idea which failed to commend itself at Rochester 7, concerned how to think about very unstable particles (that is resonances like the (3,3) or the ω). States corresponding to single stable particles were known to be associated with pole singularities on the real axis in the complex plane (p. 58). Lüders wished to describe resonances by poles at truly complex points, off the real axis. The imaginary part of the 'energy' could be associated with the lifetime in a natural way.[8] This incurred the displeasure of T.D. Lee, always a bit of a toughie in dealing with those with whom he disagreed, and he insisted, 'Unfortunately, a lifetime implies a physical measurement and not just a mathematical definition.'[9] I remember the emphasis with which he spoke. It would not be long before he, and many other people, would be looking at complex poles.

There was more sympathy for another idea floating around, which was to prove to have a future, but not in the form of its original formulation. As the collection of baryons (nucleons, Λ, Σs, Ξs) got established, people began to feel that there was really rather a lot of them. It would be theoretically agreeable if they could be assimilated to each other, differences in strangeness notwithstanding. There was a hankering for a higher symmetry than that simply provided by isospin. Gell-Mann talked of such a scheme. 'Let me suggest a name which certainly will not stick: "global symmetry" [it did not stick, neither did the particular idea itself] . . . it would be very attractive with this kind of thinking to suppose that the 8 baryons are really 8 states of one particle, the baryon.'[10] Schwinger had been thinking along similar lines. He said, perhaps with relief at having an ally this time, 'I agree with him [Gell-Mann] completely.'[11] Even Oppenheimer committed himself to a clear and straightforward statement. 'A new and startling development of the last decade is that the stronger the interaction the broader the group that it admits. I do not think we have heard the last of this or understood it at all.'[12] He, too, shared the expectation that symmetry was something capable of further exploitation.

Part of the difficulty in making any real progress along this line lay in mathematical ignorance. If lots of particles were to be combined in some way, it would require a symmetry naturally expressed mathematically through group theory. That would be a generalisation analogous to the way in which proton and neutron had been combined into nucleon states utilising the symmetry group of isospin 'rotations'. For our purpose the mathematical theory of groups can be thought of as studying the patterns formed by transformations, in this case transformations linking together particles which are in some way similar. In 1957 the only groups that theoretical physicists could think of were just direct generalisations of the familiar idea of rotations. Schwinger said, 'It is not astonishing that the only possibilities referring to internal symmetries are the successive rotation groups in Euclidean dimensions.'[13] That was not true, but no one in the physics community then knew about the more general possibility of Lie groups. This is one of the few points in our story where a few minutes conversation with a colleague in the mathematics department would instantly have opened up a new and fruitful line of thought.

There was also an idea in the air at Rochester 7 which was not to prove to have the bright future that many might have expected. This was the recurrent doubt being expressed about the adequacy of quantum field theory. Gell-Mann said, 'We all have our doubts about field theory. For most experimental physicists these doubts assume enormous proportions.' With characteristic caution he went on to say, 'I *somewhat* share their feelings.'[14] Even that arch field theorist, Schwinger, spoke only of its being 'a valid model in a certain domain .... It fails at arbitrarily small distances.'[15]

Although it did not make as big a splash as parity violation, some very interesting physics had been going on which was concerned with the properties of neutral K-mesons. Recall that K^o and its antiparticle $\bar{K}^o$ are distinct from each other. They differ in strangeness. Gell-Mann and Pais realised in 1955 that for the particle decays which did not respect strangeness, this differentiation became irrelevant. It was replaced by another consideration. Originally, Gell-Mann and Pais thought in terms of charge conjugation C, which was then believed to be conserved in weak decays, but when that dropped out it could be replaced by CP. The latter also turns K^o into $\bar{K}^o$, and vice versa. If the weak interactions conserve CP to a high degree (which they do), the decays must involve states with definite CP properties. These states are neither K nor $\bar{K}$ but are the combinations

$$K_1^o = \frac{K^o - \bar{K}^o}{\sqrt{2}},$$

$$K_2^o = \frac{K^o + \bar{K}^o}{\sqrt{2}}, \qquad\qquad [1]$$

which are, respectively, even and odd under CP. For a while, Gell-Mann and Pais hesitated because of the unusual character of such 'particle mixture' states as represented by [1]. However, the states [1] are real possibilities because of quantum theory's fundamental superposition principle which permits unexpected mixtures.[16] The upshot was that the neutral Ks were predicted to have two different sets of decay modes, with differing lifetimes: one associated with K_1^o and the other with K_2^o, and neither of definite strangeness. This was indeed found to be the case, with the lifetime of K_2^o being 800 times longer than the lifetime of K_1^o. Pais and Piccioni went on to elaborate some further extraordinary consequences. Suppose one started with a beam of K^os, which had positive strangeness. After a time long with respect to the K_1^o lifetime, but short with respect to the K_2^o lifetime, the beam would have turned into K_2^os. Consequently, it would then, in its strong interactions in flight, exhibit the presence of both positive and negative strangeness. Passing it through matter would leach out the negative strangeness (through interactions producing hyperons) and restore the original positive strangeness in an act of regeneration. These amazing and amusing predictions were amply confirmed, as Alvarez was able to report to Rochester 7.

Alvarez was also able to say, 'Every year at Rochester there is a compilation of the latest data on the various strange particles, masses, lifetimes and such things.'[17] These compilations were the origin of the celebrated 'Particle Properties Data Booklets', produced every year since, carried in their pockets by all high-energy physicists, and often the most substantial link between the worlds of theory and experiment.

8 *Geneva 8*

The international character of the Rochester meetings had steadily been increasing over the years. By 1958 the time had come to give this explicit recognition. From now on the Conferences would be held under the auspices of the International Union of Pure and Applied Physics, and they would go on their travels around the world. At the same time a small but significant change took place in their title. The word 'Nuclear' dropped out and in future the conferences were to be concerned simply with 'High Energy Physics'.

The first of these new-style meetings was held in Geneva, from the 30th June to the 5th July 1958. The choice of location was appropriate since it linked the conference with CERN, although the actual sessions were held in the University of Geneva. Attendance on this occasion was limited to 300, but special and successful efforts were made to compensate for this restriction by producing prompt and extensive Proceedings. For the first time they were properly bound in hard covers and the volume has a thoroughly professional air about it, with many clear and useful graphs and figures, handsomely reproduced. The sessions were also reorganised and for the first time systematic use was made of rapporteurs to summarise material from different sources. Generous time was allowed for discussion following these rapporteur talks. Certain stock phrases now begin to make their appearance in the Proceedings. From the rapporteurs: 'I must apologise for not being able to include in my report all the contributions I have received.'[1] From the floor: 'for the sake of completeness',[2] as someone seeks to insert, under the cloak of discussion, his pet idea which got omitted from the report. Writing about the Conference in *Physics Today* I said of rapporteurs, 'Clearly one of their greatest problems was finding the balance between the complementary requirements of adequately reporting any one new idea or experiment and of doing justice to all the new ideas and experiments that the year had produced.'[3] The worst rapporteurs were those who appeared unwilling to back their own judgement and so crowded everything in, to produce an indigestible talk, narrated at high speed as slides or transparencies, crammed with information, flashed on and off the screen before one could begin to take them in. Something of the cosy family atmosphere of the old-style Rochester had been lost. I commented at the time that 'Geneva did not seem to encourage informal discussions outside the conference sessions quite to the extent that Rochester used to.'[4]

The most extensive progress made in the year had been in the area of weak interactions. Virtually all the old results on beta decay were found to be wrong, and an entirely new picture emerged. Putting the matter more technically, in a terminology which will inevitably seem a little opaque to the non-professional reader, it had previously been asserted that the interactions were of S and T type; now it was known that they were V and A.[5] This gave great pleasure to the theorists, for the $V-A$ picture was susceptible to a particularly attractive interpretation. I shall have to describe it in slightly gnomic terms since a more articulate account would call for more technical detail than I am permitting myself to employ. The closest one can get to extending the two-component idea (p. 63) to massive particles is a requirement called γ_s-invariance, or chirality (a somewhat learned word for the maximal attainable degree of handedness). The imposition of this requirement uniquely picks out V and A. The idea had occurred more or less simultaneously to Marshak and Sudarshan, to Feynman and Gell-Mann, and to Sakurai. When the time is ripe for a discovery such coincident thoughts are by no means uncommon, particularly in high-energy physics where at any one stage of development of the subject a lot of clever people are focussing their attention on a comparatively restricted and well-defined set of problems. Human nature being what it is, this can sometimes lead to a certain edginess about claims to priority. I recall hearing about a later meeting discussing chirality at which such feelings were manifest. Eventually, Feynman got up and said, in his perky way, 'I would like to claim to be the last person to have thought of this idea.'

Pure chirality would lead to the combination $(V-A)$, whilst, experimentally, the result for beta decay was $(V-1.3A)$. This was not a fundamental problem. In nuclear decays the strong interactions would be expected to modify (renormalise) the coefficients. Feynman and Gell-Mann pointed out that such a renormalisation effect did not appear, however, to make any difference as far as the vector part V was concerned. The coefficients measured in beta decay (with strong interactions present) and in muon decay (with strong interactions absent) were closely similar. It was later realised that this coincidence afforded an important insight into the nature of the currents involved in weak interactions.

Only one major difficulty remained for $V-A$ theory. It related to the decay of the pion. Almost all the time the charged pions decay into a muon and neutrino. Only very occasionally is the muon replaced by an electron to give what one could call pion beta decay. The smallness of the ratio of these decays is well understood in terms of chiral symmetry. It arises from the smallness of the electron's mass in relation to the mass of the muon. This implies that the former is much closer to being in a state of definite handedness than the latter. (Remember only massless particles can be strictly single-handed.) If the electron were purely left-handed, pion beta decay would be forbidden altogether.[6] As it is, it is constrained to be very small. Feynman gave a vigorous account of this at Geneva, with impressive manual illustration of the handednesses involved. The problem that was worrying people was a quantitative one. The ratio that was predicted was very small (10^{-4}) but experiment at the time seemed to be suggesting that it was even smaller, by as much as an order of magnitude. It proved to be yet another case of the peculiarity of weak interaction measurements. In due course the results settled down to the value predicted by $V-A$.

In that expression V stands for vector and A for axial vector. This implies that the weak interactions involve currents (which have a vectorial character), not altogether dissimilar to the electromagnetic current (which is a particular sort of V). Electromagnetism is mediated by the exchange of a boson, the photon. From this time onwards, speculation recurs that the weak interactions might be similarly mediated by one or more particles of a type given the generic name of 'intermediate vector boson' (IVB). Like the photon, IVBs would have spin 1. Such particles would also have to differ from the photon in a number of ways. For instance, the weak interactions involve changes of charge (the neutron decays into a proton) and so a theory of IVBs would need to include the possibility of some at least being charged, in contrast to the neutral photon. Secondly, the very short range of the weak interactions implies that the IVBs must be very massive, in contrast to the photon's masslessness. It was soon realised that this latter fact would permit some degree of assimilation of the IVBs and the photon to each other, at least as far as the intrinsic coupling strengths were concerned. Weak interactions are, of course, very much less strong than electromagnetic interactions at ordinary energies, but this could be due to supression effects operating at energies much less than the IVB mass. It was conceivable that at IVB energies the two interactions might prove to be of comparable strength. There remained the difference that electromagnetism conserves parity and weak interactions do not. At Geneva 8 these ideas were not much more than glints in the eyes of speculatively minded theorists. In his summary at the end of the Conference, Oppenheimer said, 'We have heard arguments for and against an intermediate boson; and surely the search for the effects of it . . . will be one of the things for which we will all be waiting.'[7] It was to prove a long wait.

An interesting development had taken place which exploited the relative smallness of the pion mass (0.15 on the scale of the nucleon mass). Goldberger and Treiman had derived a simple relation between constants relating to the strong interactions and the axial current (A) of weak decays. Their original argument was in the nature of inspired trickery. The Goldberger−Treiman relation is now understood as a good approximation resulting from the relative smallness of the pion mass − a sort of low-energy theorem (technically: PCAC − partially conserved axial current; more about this will be given later).

The rapporteur dealing with advances in theoretical understanding of strong interactions was Geoffrey Chew. Geneva 8 was the beginning of a period in which Chew was to play a major role at a succession of Conferences, as the identifier and salesman of ideas. I have written elsewhere that 'his talks were always eagerly awaited because of their inspirational and encouraging tone which helped to sustain one's possibly flagging spirits, and also because of his ability to put his finger on whatever was most promising in the year's crop of ideas.'[8]

Chew started by saying of strong interactions that 'we must be even more liberal than usual in our definition of a "successful theory". I think we must call a contribution successful if it leads to any correlation between experiments which has not previously been recognised.'[9] That was a pretty minimal definition of success, framed in a positivistic spirit of simply relating observations. The first part of Chew's talk dealt with nuclear forces. People would not give up searching at the end of the rainbow for the perfect potential. Over the years, the current candidate's ad hoc complexity had multiplied considerably. Chew characterised the favoured potential of 1958 as 'a real bastard' of 'almost unbelievably obscure parentage'.[10] He listed ten physicists who had shared in its generation.

Perhaps with relief, Chew turned to dispersion relations. He believed that a 'reasonable attitude' to them was that they 'represent a conjecture that scattering amplitudes can be extended into the complex plane with the minimal number of singularities required by the analyticity of the S-matrix'.[11] His aim was elaborated when he went on to say that what was involved was 'the attempt to use them as fundamental dynamical relations which replace the usual field equations. There is a hope that in this way one does not commit oneself to regarding particles as elementary',[12] so that one might avoid troubles then thought to be present in field theory. That was a manifesto outlining a programme that was to occupy Chew and many colleagues for years to come.

Meanwhile, the inspirational message for 1958 had been provided by the young Stanley Mandelstam. He was a South African of powerful mathematical ability who, in an obscure and difficult paper, had proposed a much more extensive use of analyticity than had been attempted before. The basic idea was this. Consider the scattering process $1 + 2 \rightarrow 3 + 4$. Let us represent the momenta as in the figure, where all are drawn as incoming.

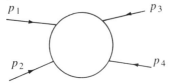

That means that we have changed the signs of p_3 and p_4 from their natural outgoing values. This has been done to make everything look as symmetrical as possible. Energy-momentum conservation implies that

$$p_1 + p_2 + p_3 + p_4 = 0.$$

We can define a quantity

$$s = (p_1 + p_2)^2 = (p_3 + p_4)^2,$$

which turns out to be just the square of the centre-of-mass energy for the process $1 + 2 \rightarrow 3 + 4$ (A). This is essentially the variable we use when writing down the 'old-fashioned' (that is, last year's) dispersion relations for this process. At the same time we fixed the momentum transfer, whose square is the variable

$$t = (p_1 + p_3)^2 = (p_2 + p_4)^2.$$

Now turn your head on one side and read the figure vertically rather than horizontally. It now looks like the process $1 + 3 \rightarrow 2 + 4$ (B), a crossed process (p. 58) for which the energy is t and the momentum transfer is s. In other words, the two variables have changed roles in reading the figure the two different ways. (The reason for writing the vectors in a symmetrical ingoing way was to facilitate this kind of thinking.) Mandelstam suggested that one should consider analyticity in s and t together. He therefore wrote down a double-dispersion relation in both variables.[13] All that may sound very abstract and mathematical, but Chew realised that it held promise of physical progress, not least because it tied together the two processes, (A) and its crossed partner (B), in a single analytic function. Some pretty clever and restrictive conditions would be required to ensure that this function was consistent with unitarity whichever channel (process) it was applied to. That is, it had to conserve probability and make physical sense whether one used it for process (A) or for process (B). Chew tried to persuade his audience of the importance of Mandelstam's conjecture. 'This work will perhaps seem baffling to experimenters, but I assure you that we shall quickly come to important practical results.' He went on to say that Mandelstam had 'an iterative procedure which he hopes will allow a numerical solution'.[14] That was an optimism which was not to prove justified.

At a more theoretical level these ideas led to the notion of bootstrappery — the breathtaking concept that there were no privileged fundamental particles but that everything was made out of everything; that the physical world lifted itself into being like a man levitating by pulling on his own bootstraps. This possibility arose from crossing's ability to read physics in two directions. Take the ρ meson, a metastable state of two pions. In one channel (s, say) it made its appearance as a resonance pole. But it would also figure in the crossed channel (t), where it represented an exchanged system which, in classic Yukawa fashion, generated a force between the pions. Just this force could bind the pions together to give the ρ resonance! One had the possibility to defy the classical logicians, to identify cause and effect, to claim that it was the ρ that made the ρ. Clearly, some delicate consistency conditions would be required to bring this about satisfactorily. Clearly, if one generalised the idea so as to make everything out of everything, these conditions became immensely more delicate, extensive and specifying. It was the grand, and grandiose, hope that a uniquely determinate theory would result. Alas, bootstrappery has come to naught, not only because it is impossibly difficult to implement but also because it is highly plausible, as we shall see, that there are specific constituents making up the variety of observed particles.

Despite Chew's enthusiasm, the Mandelstam representation (as the double-dispersion relations were called) did not cause an immediate stir. It was too unfamiliar. In the discussion, the Russian, N.N. Bogolyubov, equally powerful in mathematical physics and Soviet scientific politics, asked if it were proved or just an ansatz (a trial guess). It was, of course, the latter.

A background puzzle to all this was the difficulty in understanding the precise physical basis for assumptions of analyticity properties. These were now far outstripping anything that could be proved rigorously on the basis of local field theory. The nuclear physicists had been thinking longer about such questions and one of them, Aage Bohr, son of the famous Niels, was invited to share his insights with the Conference. They did not seem to help with the problem of unphysical regions (p. 59). From the floor, the Swiss theorist, Res Jost, uttered the impressive warning that those regions were 'as bad and treacherous as any point in the complex plane'.[15]

One chink of light was shed on the problem of high-energy behaviour (p. 59). It was reported to the Conference that the Russian theorist, Ya Pomeranchuk, had shown that, under very general conditions, the total cross-sections for particle and for antiparticle scattering become the same at very high-energy.[16] This result is now universally known as the Pomeranchuk theorem.

The last refuge of idiosyncracy and anarchy was the session on Fundamental Theoretical Ideas. This was an area which did not lend itself

'No credits for the future' (Wolfgang Pauli) .

to the bland treatment of the rapporteur. In such speculative matters the individual still insisted on his right to speak. Pauli was the chairman and he opened the proceedings with some characteristically acerbic remarks. 'This session is called "fundamental ideas" in field theory, but you will soon find out or have already found out that there are no new fundamental ideas ... you will also see that there are two kinds of ignorance, the rigorous ignorance or the more clumsy ignorance. You will also hear that many speakers will want to form new credits for the future. I am personally not willing to give such credits ....'[17] It would be no use waving your hands in front of him and expressing the hope that it would all work out right in the end.

The speaker who incurred Pauli's greatest wrath was none other than Heisenberg. He had conjectured a 'non-linear spinor equation', whose solutions he thought would correspond to the structure of matter as it was then known. Not only was his equation hard to work with, but in the course of the attempt use was made of the dangerous concept of an indefinite metric, something which could result in the appearance of unphysical ghosts (p. 60). What added fuel to the confrontational fire was the fact that for a while Pauli had been persuaded to go along with his old friend's ideas. He was now in the frame of mind of a man who has parted with his £100 but finds that he has not bought London Bridge after all. Heisenberg was continually interrupted by Pauli, who wagged a reproving finger whilst uttering such remarks as 'mathematically objectionable', 'This I discussed [and rejected] in April and I wonder that you again repeat it all' and, most frequently, 'no credits for the future'.[18] It was a scene at once farcical and sad. Justification lay with the sceptical Pauli but Heisenberg was one of the greatest physicists of the twentieth century who should have been able to enjoy a more dignified close to his career.

The Conference received the first results from the synchrotron built at Dubna, outside Moscow. There had not previously been any high-energy accelerator physics in the USSR and the Russians had decided to make a grand entry right at the top, with a machine whose 9 Gev energy was the highest in the world at the time. It gradually became clear that they had been too ambitious. The machine did not work well, though I remember some years later being impressed by its control console, panelled with rare hardwoods in that Edwardian taste which seemed to commend itself to communist fancy, whether in hotels, aeroplanes or synchrotrons. Meanwhile, for ultra-high energies the cosmic ray people still held the field. Piccioni reported that, 'There exist, the world over, somewhat more than 50 events in emulsion with primary energy greater than 10^{12} ev.'[19] It is only now and in the near future that accelerators will begin to explore that regime.

9 *Kiev 9*

By 1959 the Rochester Conference had arrived in Russia. The Ninth International Conference on High Energy Physics was held in Kiev, from the 15th to the 25th of July 1959. The city had been extensively damaged during the war and was rebuilt in the baroque curlicue style approved by Stalin.

I travelled out in a party headed by Abdus Salam. It was a leisurely 'plane journey and we stopped off in Prague and Budapest on the way. The latter was still a city somewhat under siege after the 1956 uprising. Salam was a foreign member of the Hungarian Academy of Science and this smoothed our way. We were wafted into the centre of the city by limousine and shown the sights, which included a swimming pool with a wavemaker, which the hotel manager told us had greatly pleased Edward, Prince of Wales, on a prewar visit. The gratification induced by all this increased when we learnt next day that other British colleagues had not been so fortunate. They had been forced to spend the night behind barbed wire in the closely guarded transit compound at the airport.

The Conference was conducted in English and Russian. Rapporteur talks were simultaneously translated. Reviewing the Conference in *Physics Today* I wrote tactfully, 'Simultaneous translation is a difficult art at the best of times and when it is concerned with talks using a highly specialised vocabulary it is not surprising if it does not always achieve perfection.'[1] The ill effects were considerably mitigated by the heroic labours of rapporteurs and scientific secretaries who produced duplicated translations of draft scripts before most of the talks were delivered. More seriously disrupting was the need to pause for consecutive translations of contributions to the discussions. This effectively destroyed the cut-and-thrust of debate. Outside the sessions, however, the fact that we were all housed in the same hotel, together with the leisurely character of Russian restaurant service, provided many opportunities for informal talks in small groups.

For the organisers, the Conference itself is only half the battle. There remains the task of producing its Proceedings. Kiev slipped back several notches in this respect compared with Geneva. The Proceedings reverted to a poor-quality typescript and they took an unconscionably long time to appear. They also ran into two volumes, totalling more than a thousand pages.

Mandelstam's double-dispersion relations were now all the rage. A down-to-earth experimentalist like Alvarez could acknowledge that what had seemed an obscure theoretical novelty had become 'the bread and butter of the experimentalists'. He regarded such rapid transfer of ideas as providing 'a complete justification for our annual conferences'.[2] This had come about through the recognition that analytic continuation could provide access to information about scattering processes which were impossible to study directly in the laboratory. The basic trick can be illustrated by considering the scattering of pions by pions. No experimentalist had a pion target. You could not get enough of them together for long enough to have a chance of hitting anything. Pion–pion scattering could not be investigated in any straightforward way. Mandelstam, however, had taught us to consider treating the momentum transfer (t) in pion–nucleon scattering as a complex variable, extendable (continuable, a mathematician would say) from its physical values. The simplest singularity in t which you would encounter in making the continuation would be a pole, corresponding to a single pion state. Poles are singularities which stick out like sore thumbs if you are at all near them (the function behaves in their neighbourhood like $1/x$ near $x = 0$). Poles are therefore relatively easy to pin down in mathematical terms. The 'residue' at the pole (the coefficient multiplying the $1/x$ behaviour) in the case of the process

$$\pi + p \rightarrow \pi + \pi + p \qquad [1]$$

is proportional to the otherwise unknowable pion–pion scattering amplitude. The flavour of the argument is indicated diagramatically in the figure. The reader for whom the mathematical details are elusive must content himself with assent to a popular slogan of the time. Analytic continuation to pole singularities enabled one to 'do scattering with unstable targets'. Obviously, that is an exciting prospect.

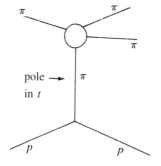

Equally obviously it is not as easy as all that to accomplish. The practicalities proved pretty tricky to implement. Commenting on this 'new branch of mathematical physics which goes in some quarters under the distasteful name of "polology" ', I said, 'Analytic continuation by computer is, to say the least, a hazardous process, particularly when data contain large errors.'[3] Some of the claims for results obtained by these new methods (such as the disturbing possibility that charged and neutral K-mesons, despite being members of the same isospin doublet and so expected to be closely similar, might have opposite parities) were to prove to be rash and wrong.

Chew summarised the basic theoretical situation as it appeared in 1959. The double-dispersion relation approach 'may conservatively be regarded as a generalisation of the effective range approximation [that is, a low-energy calculational device of real but limited value] ... it may also be regarded as something very deep and fundamental'. In view of his later attachment to S-matrix ideology, it is interesting that Chew went on to say, 'In either case, certain concepts of relativistic local field theory are involved in an essential way.'[4] With characteristic optimism he concluded, 'If the Mandelstam rules can be accepted, an enormously powerful approach to strong-coupling physics appears to be opened.'[5] Many man-years of labour were to be expended at Berkeley before the hollowness of that promise became apparent.

All this time there remained the fundamental theoretical question of whether Mandelstam's inspired guess was actually right. Was the singularity structure in s and t as simple as he had supposed? I commented in my report that, 'The problem is so difficult that it has had the effect of resurrecting perturbation theory from the waste paper basket.'[6] Feynman integrals define analytic functions. The study of the singularities they possess might reasonably, if not rigorously, be expected to shed light on what to expect more generally. Used in this way, perturbation theory was a heuristic device for the exploration of the formal properties of relativistic quantum theory, without any unrealistic commitment to its being a sensible calculational procedure for use in strong-interaction physics. Feynman integrals are associated with Feynman diagrams: pictorial representation of processes proceeding by the exchange of 'particles'. Inverted commas are called for since the 'particles' involved internally in the diagrams (that is to say, just exchanged and not being particles in either the initial or final state of the scattering) are what are called virtual particles. They do not have to correspond to physical masses in the way required of the external particles.[7] The more complicated is the diagram, the less perspicuous are the properties of the corresponding integral. It was natural to start with the simplest cases: the lowest-order diagrams as we say. Some of these had been looked at by some East-coast Americans: Karplus,

Sommerfield and Wickman. Their conclusion was that in certain circumstances Mandelstam's conjecture would not work. Its failure was associated with the occurrence of the same 'anomalous thresholds' which had proved the downfall of Schwinger's simplistic ideas. Karplus, Sommerfield and Wickman were also able to suggest a certain geometrical test (technical aside: the construction of a dual diagram) which would indicate when this trouble might be expected. This duality notion had been generalised by my friend and contemporary John C. Taylor in England. These ideas burst on the Kiev scene in fully fledged form in a colourful exposition by Lev Davidovitch Landau, who had also been thinking about the problem.

All physics conferences have their edgy moments when people with differing styles and views oppose each other. At Kiev 9 the most dramatic confrontations involved two Russians of very contrasting temperaments and circumstances. One was Landau − brilliantly intuitive, Jewish and at odds with the regime because of his liberal-minded outspokeness. The other was N.N. Bogolyubov − profoundly mathematical, a communist and the powerful Secretary of the Soviet Academy of Science. Both had made important contributions to theoretical physics. One was a man of ideas, the other was a man of proofs. In Russian style, each was surrounded by his group of loyal supporters. In his opening address to the conference Blokhintsev had said, 'In our time scientists of various countries understand each other perfectly when they discuss their professional matter.'[8] Ironically, the most obvious failure of minds to meet at Kiev 9 involved these two Russians.

Landau produced a set of equations which he claimed determined the singularities of scattering amplitudes. They included the dual diagrams as a special case in appropriate circumstances, but they were more general than that. The Landau equations had about them an intuitively appealing character which convinced one that here was something significant and correct. But where did they come from? They looked as though they might be derivable from perturbation theory but Landau insisted emphatically that this was not the way to think. He believed that quantum field theory was plagued by ghosts and he consequently placed no reliance on it. He was, in fact, putting forward his equations as field theory's successor. 'If we speak of the derivation of this technique from conventional theory, this can be understood only in the sense in which quantum theory was derived from the correspondence principle some time ago, and this derivation has no bearing on the long deceased perturbation theory.'[9] The rather grandiose claim was being made that just as Newtonian theory would not do to discuss atomic phenomena but, of course, whatever did describe atoms had to fit on smoothly to Newton's account of the motions of larger bodies, so quantum field theory would not do for elementary particle physics, although Landau's new ideas were acknowledged to bear some consonant relation to its way of talking. For most of us this was baffling but exciting. For Bogolyubov it was clearly infuriating. The question of singularities was a mathematical question and he, after all, was a very skilful mathematician. He added a note in proof to the Proceedings which said, 'I wish to stress that the problems under discussion [in connection with analyticity and dispersion relations] are purely mathematical in character. ... Because of the mathematical character the physical intuition may not work.'[10] It was not hard to guess whose intuition he had in mind.

But the charm of theoretical physics lies in its blending of mathematical argument with intuitive discernment. Landau's insight had enabled him to penetrate to the heart of the matter. As a distinctly non-establishment man he was not given much time by the authorities to develop his ideas in the conference proper. The printed Proceedings do not even contain the Landau equations written out explicitly. Yet the quest of physicists for understanding is not to be frustrated by the rigidities of the system. An informal session was hastily organised at which Landau was given a full opportunity to present his work. Attending it was one of the most formative and exciting experiences of my life as a theoretical physicist. I certainly did not understand what Landau was driving at, but I went away determined to try to find out. Fairly soon, working with Gordon Screaton, I was able to see how Landau's equations could indeed be derived from perturbation theory's Feynman integrals. It was somewhat later that I began to understand that the equations could also be understood in a more general way than that. Analyticity and unitarity (the need to make physical sense by conserving probability) together form a very powerful constraint when combined with crossing's requirement that a single amplitude be capable of describing several different processes. The Landau equations represent the minimal singularity structure which is consistent with that complicated interlocking set of conditions. It was, indeed, a mathematical question to figure out that this was so. What is breathtaking to behold, and commands the highest intellectual admiration, is the profound intuition which enabled Landau to write down the answer, complete and fully formed.

There is a curious footnote to all this. In his talk Chew remarked, 'Parenthetically, I may pass on the statement given me two days ago by Symanzik that Regge has achieved a firm proof of the new two-dimensional spectral representation [the Mandelstam representation] for amplitudes generated by a superposition of Yukawa potentials.'[11] That sounded amusing rather than significant. After all, potentials are really a non-relativistic concept and in this domain of quintessentially relativistic physics one might say, 'So what?' Actually the work of Tullio Regge was to prove another of those clouds no larger than a man's hand which soon turn into a torrential downpour. It was not what he had done, but the way that he had done it, which was to prove extremely fruitful. Regge had had the clever idea of turning angular momentum into a complex variable. That was by no means an obvious trick to use. Angular momentum physically is a discrete variable (it takes whole number values, 0, 1, 2, . . .) and its extension into the complex plane is a much more subtle business than that of a continuous variable like energy (which can take a whole range of values, so that one has a stretch of the real axis from which to start). More of this anon.

The complex plane apart, there was a bread-and-butter, rather than caviare, taste to a good deal of Kiev 9. Rapporteurs protested at the last-minute character of many of the contributions they were called upon to report. Jack Steinberger, who was covering certain aspects of strange-particle physics, received only three papers but, 'As usual they were given me at the last minute and it was not possible to digest them properly. All blame for a thoughtless report rests therefore squarely on the contributors.'[12] Weak interaction was settling down into a triumph for $V - A$. Even the pion had fallen into line. Ingenious discussion of why it did not beta decay had been curtailed by the discovery that it did. Straws began to blow about in the wind, which were to prove to be tokens of further advances yet to come in that perennially interesting field of physics. It was noted at Kiev 9 that hyperon beta decays seemed to be an order of magnitude weaker than those of nucleons. This would only cease to be puzzling when people began to understand more about the currents involved in weak decays. In a characteristically perceptive and obscure communication, Nambu suggested, 'there seems to be a possibility of learning something about elementary particle physics by studying superconductivity.'[13] The immensely important concept of spontaneous symmetry breaking was beginning to make an oblique entry into the subject. The central idea is that solutions of a theory may possess less symmetry than would seem to be implied by its basic structure. As a very simple example, consider a pin balanced on its point. The system has symmetry about a vertical axis but its lowest energy state corresponds to the pin's having fallen flat in some direction. That selection of a specific horizontal direction breaks the symmetry about the vertical. The direction in which it does so is arbitrary, being triggered by the infinitesimal fluctuation which spontaneously causes the pin to tip over in that particular way. The idea that the lowest energy state (the vacuum) might be displaced in a symmetry-breaking way opens up the possibility that field theories which at face value might seem to describe massless particles, in fact yield massive particles in their actual physical states.

The possibility of unifying electromagnetic and weak interactions continued to attract attention. In discussion, the swarthy Russian theorist Zeldovitch acknowledged that, 'There are extremely interesting and bold suggestions by Salam and Ward, Glashow and remarks by Gell-Mann . . . .

In this case an organic fusion of the theories of electromagnetism and weak interactions becomes possible. However, this requires $M_X = 40\ M_{nucleon}$.'[14] Such a mass for the intermediate vector boson seemed enormous and inaccessibly remote in those days.

The Kiev Conference saw Bruno Pontecorvo meeting for the first time for some years with his former British colleagues. Although never accused himself, he had left Britain suddenly at a time of atomic spy scandals and had settled in the Soviet Union. He had not been wasting his time there for he drew the Conference's attention to an interesting apparent anomaly in the behaviour of weak decays. The normal decay of the muon is

$$\mu \rightarrow e + \nu + \bar{\nu}. \tag{2}$$

One might expect that sometimes the neutrino and antineutrino would 'eat each other', that is, produce in their stead a gamma ray, giving the decay

$$\mu \rightarrow e + \gamma. \tag{3}$$

In an intermediate vector boson theory of weak interactions one could estimate this effect with reasonable plausibility. It was expected to occur about once in a thousand decays. Yet it was known to be absent to an accuracy of about two in a million. Pontecorvo suggested that the neutrino associated with the muon and the neutrino associated with the electron might belong to different species (that is, that in [2] we have ν_μ and $\bar{\nu}_e$, respectively) and therefore be incapable of annihilating each other to give [3]. At the time that sounded a pretty bold, not to say wild, speculation. The idea did not commend itself to Marshak, who found it 'quite inelegant'.[15]

It had become customary for each new accelerator to succeed in discovering its own new particle. The people manning the Russian synchrotron at Dubna were obviously anxious to follow in this tradition. They reported at Kiev 9 some thinnish evidence for a heavy meson of double strangeness. Alas, it was part of the rather dreary track record of that machine that the 'dubnon' did not stay the course but soon vanished from the list of putative particles.

The end-of-Conference summary was given by the veteran Russian theorist, Igor Tamm. One of my warmest recollections of Kiev 9 is of the evident affection and respect which he drew from all who knew him, whether from East or West. This was due, not simply to his considerable theoretical abilities, or even to his evident friendliness and modesty, but also because of the courage and integrity he had shown in the difficult Stalinist years. He took a rather sober view of the achievements of the year just past. 'No particular sensitive or surprising scientific news was revealed.'[16] He drew attention to the increasing sophistication of experiment, indicated by the construction of large bubble chambers and the devising of automated data procedures, and expressed himself as being 'puzzled by the lack of proportion between the money spent on accelerators, on the one hand, and auxiliary devices on the other'.[17] That had certainly been so in the past, when the policy had often been to build the machine and then begin to think what to do with it afterwards. The years ahead were to redress the balance.

After the Kiev Conference a great party of us flew on to Moscow for a few days sightseeing. We gathered at the airport, which I recall (surely incorrectly?) as being a grass field with a marquee, just outside the city. There was some consternation as we prepared to board the plane. It seemed that Stanley Mandelstam, whose name had been on all our lips, had not been provided with a valid ticket for the journey. In those days, Intourist's answer to every problem was, 'Is impossible'. Were we, then, to leave the inventor of the double-dispersion relation behind? An honest American experimentalist with the build of a footballer, Dave Frisch, said in a wave of warm emotion, 'Boys, if Stanley doesn't go, we all don't go.' One of his more sophisticated and cautious theoretical colleagues said, 'Wait a minute.' For a little while it was a tense scene. Fortunately, it was not long before the impossible turned out to be possible after all. Stanley made the trip with the rest of us.

10 *Rochester 10*

In 1960 the Conference returned to its birthplace. About 300 participants from about 30 countries assembled at Rochester during the period 25th of August to the 1st of September. Yet another variation in procedure was tried out. The time available was divided into three. One third was spent on parallel working sessions. These were really old-style 'mini-Rochesters', devoted to specific topics and affording individuals the opportunity to speak of their own work. They proved as resistant as ever to neat organisation and exact time keeping. There followed two days of rapporteur summaries and evaluations of these sessions, with a final period in which certain invited plenary speakers were given the freedom of the floor. The burgeoning state of the subject meant that, even with these arrangements, there was insufficient opportunity for all to speak who wanted to do so. The Conference Proceedings print details of tens of papers submitted to various sessions but not presented due to lack of time.

The report on Rochester 10 in *Physics Today* was written by Mike Moravcsik. (Two metres tall, Gell-Mann used to refer to him as 'the greatest living physicist', a joke which no doubt gained added edge from Murray's obvious grounds for contesting the title, less literally interpreted.) In a retrospective glance at the previous decade, Moravcsik noted that the energy available in the laboratory had increased by a factor of 100. High-energy physics was big business, so much so that it 'has become the concern of governments and its direction is part of national policy'.[1] Moravcsik did not take an altogether rosy view of progress. 'The efforts to penetrate into higher and higher energy regimes do not stem from an understanding of phenomena at lower energies. In this sense, perhaps, high-energy physics is moving too fast.'[2] To some extent that judgement reflected his own interests, which were of a bread-and-butter kind, concerned with the detailed exploration of phenomena. It failed to take into account the possibility that extreme regimes, such as very high energies, may by their special character exhibit simplicities which are pointers to an underlying structure, more difficult to discern in

86

ordinary circumstances — just as people's true characters are often only revealed in situations of unusual stress. The future was certainly to lie with the pursuit of simplicity through extremity.

Governments did not only impinge on high-energy physics through their funding activities. At Rochester 10 there were still problems with the McCarran Act for some of those invited. Also a number of outstanding Soviet physicists cancelled their participation at the last moment. Reasons were not given, but it was commonly assumed that the authorities at home would not let them out of the country.

I arrived at Rochester in a quivery state. During the past year I had been working with John C. Taylor and Peter Landshoff, then a student of mine, on the problem of trying to establish the truth of the Mandelstam representation for *all* the Feynman diagrams relating to a given scattering process, in the circumstances in which it was known to be true for the lowest order diagrams (absence of anomalous thresholds). This would obviously be a worthwhile result if correct. It would not amount to a proof which would satisfy a mathematically minded person like Bogolyubov (after all, one did not suppose that the perturbation series converged for strong-interaction physics), but it would certainly provide considerably enhanced intuitive backing for the whole idea. Taylor, Landshoff and I thought that we had done that. Our discussion was complicated, essentially because one had to establish a very tight control which stopped new singularities 'popping up' out of the cuts due to the ones (normal thresholds) already there. It was to be my job to present this work at Rochester 10. I felt a good deal of reputation depended on how I did it. Just as I arrived at Rochester I learnt that another Cambridge colleague of mine, Richard Eden, who had done important early work on singularities and who had been away on leave, had also produced a proof. His argument seemed much simpler but I felt that this plain speaking must have been purchased at the cost of some oversimplification somewhere. The situation just appeared to be intrinsically complicated. Richard's approach was so differently expressed from ours that I could not quickly put my finger on the doubtful point in the time available. Symanzik, the rapporteur, delivered his verdict: 'Eden's technique for proving the Mandelstam representation is more perspicuous.'[3]

I now think that I could identify a gap in Richard's argument which for its filling-up would require the sort of tortuous elaboration to which we had been driven. However, that became irrelevant. When we all returned to Cambridge after Rochester 10, we worked together on these problems. In the course of that work we encountered a further difficulty of which we had all been unaware. The curves of possible singularity had been assumed to be well-behaved geometrical figures. In our joint work we discovered that

The eight Nobel laureates who attended Rochester 10. (From left to right: Segré, Yang, Chamberlain, Lee, McMillan, Anderson, Rabi, Heisenberg).

this 'reasonable' assumption did not, in fact, hold. The Landau curves were often degenerate, with geometrically peculiar properties (technically, nodes and cusps). This frustrated the sort of continuity arguments we had been employing and wrecked the proof. Later on another difficulty emerged.

Basic to all this work were Landau's equations for the possible position of singularities. By this time they had been set on a firm footing in relation to Feynman diagrams. The singularities we had been attempting to understand all corresponded to straightforward solutions of these equations. A group of us went on to discover that the equations also possessed some much less transparently intuitive solutions which could yield a yet further class of possible singularities. These 'second-type' singularities, as we called them, also proved unamenable to control and elimination by the methods at our disposal. The problem remains unsolved.

Rochester 10 saw an important addition to our general understanding of singularity structure. It was due to Dick Cutkosky from Pittsburgh. Most singularities in the complex plane are branch points, that is, if you encircle them to return ostensibly to the same point, the function does not return to the same value. An example is the square-root function $\sqrt{x}$, which has a singularity at $x = 0$. If you start at, say $x = 2$ with the value $\sqrt{2}$ and go round $x = 0$ once, you end up back at 2 with the 'other square root', $-\sqrt{2}$. Dealing adequately with branch points requires a knowledge, not only of where they are, but also of their discontinuities: how the function changes when you encircle them. For the Landau singularities, Cutkosky produced a beautiful formula for their discontinuities. Like the Landau equations themselves, it was instantly, intuitively appealing to those who feel at home in mathematical physics.

All this was exciting for those who cared about such things. Those who did not were hard to impress. Down-to-earth Moravcsik simply notes the work on the Mandelstam representation and complains that it has not been extended to more complicated processes involving the production of extra particles. (It was already known, in fact, that no such extension was straightforwardly possible.) His conclusion is characteristically wary, not to say grudging. 'It seems clear that a really basic and sweeping insight into elementary particle theory, comparable to that of the formulation of quantum theory in atomic and nuclear phenomena, is yet to come if it will come at all.'[4] Progress, as and when it came, was not made in any revolutionary fashion comparable to Newton's giving way to Heisenberg, but in a thorough and patient exploration of relativistic quantum theory.

The working session that considered all these issues was chaired by Oppenheimer. It dragged on and on and eventually it was driven to reconvening after supper. We had already been hours on the job and I felt a spurt of irritation when Oppenheimer opened that evening session by saying, 'It is good to see you here because I know that you have all come because you love physics'. I thought to myself, 'We are all here because you are an indulgent and ineffective chairman who has allowed this session to get out of hand.' I have to confess that I never really cared for Robert Oppenheimer. He had clearly been an inspiring teacher in his prewar days at Berkeley. By the time I first met him he was battered by his treatment at the hands of the AEC over security matters. The need to retain his authority and standing in another sphere was doubtless part of the reason for his mandarin manner. Perhaps there was also a further cause. When he gave the Reith Lectures on the BBC the _Observer_

'Oppy' in reflective mood.

newspaper published a profile of him. I recall its saying that Oppenheimer's secret sorrow was that he had not made a fundamental discovery in physics. By the very highest standards — and Oppenheimer would never have deigned to apply to himself standards less than those — that was true. I think he had that secret sorrow and that it drove him to attempt always to assert a superiority over those ordinary physicists with whom he came in contact. My most frequent encounters with him were a few years later when I spent a semester as a visitor at the Institute for Advanced Study at Princeton, of which he was the director. I came to feel that his gnomic way of speaking, with its stream of epigrams, phrased to be maximally striking but minimally clarifying, was a device to put the listener at a disadvantage. He was a most uncomfortable man to be with.

In the world of practical men, the singularities in the complex plane cast a shadow in the form of peripheral dynamics. Single-particle poles (p. 58) were outside the region physically accessible to measurement, but when a momentum transfer (t) was small the poles were not too far away. Their proximity made the scattering amplitudes correspondingly large. As a result, scatterings with not much change of momentum (grazing or peripheral collisions) were enhanced and they accounted for much of what was going on at high energies in strong interactions. A lot of calculation and analysis attempted to exploit this fact.

Weak interactions continued to progress. Once again, a timely idea was in the air which several groups of people got on to at about the same time. The lack of the renormalisation of the vector current by strong interactions (p. 71) was recognised as being explicable if V were a conserved vector current. (A conserved current is one which does not lose anything of what it is carrying.) That would make it analogous to the electromagnetic current, where electric charge is conserved and a similar absence of renormalisation is also evident. The electric charge of the proton (which has strong interactions) is the same as the charge of the positron (which does not). Feynman was one of those who had noted this way of explaining an otherwise odd coincidence. Other analogies could also be exploited, leading to the coining of the phrase 'weak magnetism'. Feynman was fairly reserved in his assessment of the idea at Rochester 10. 'It is just a good guess. If it goes down, nothing goes with it. ... Nevertheless it is fun to guess, so let us see how it works.'[5] His collaborator in the matter, Gell-Mann, was also wary. 'I would like to talk about the sort of structure which will crumble when the conserved vector current hypothesis is exploded.'[6] There was to be no such disintegration. It was the beginning of a good deal of thought about the structure of weak currents in which Murray Gell-Mann was to play the leading role.

People continued to toy with the idea of an intermediate vector boson. Goldberger (speaking in place of the absent Russian theorist Okun, who had been unable to come) pointed out that a heavy vector boson would give, according to current wisdom, a non-renormalisable theory. 'Whether one worries about the intermediate boson theory for this reason is a matter of religion.'[7] Later on there would be a good deal of conversion to that religion of renormalisability. One of its sacred texts was adumbrated in a remark of Nambu's. He continued to testify about spontaneous symmetry breaking, saying that 'solid state physics can provide us with useful models which help us to understand the dynamics of elementary particles'.[8] For a while yet he was to remain a prophetic voice crying in the wilderness.

Results were reported from CERN of a very accurate measurement of the anomalous magnetic moment of the muon. These $(g-2)$ experiments, as they were called, were to demonstrate over the years an exquisite exactitude, matched only by the equally precise calculations of quantum electrodynamics, in perfect agreement with them.

Great staying power was being manifested by the $|\Delta I| = \frac{1}{2}$ rule, whose 'spectacular success' after some previous apparent problems was declared by Mel Schwartz to be the 'most interesting news of the year'.[9]

Rochester 10 received a few drops of what was to become a torrent. Alvarez reported the discovery of a hyperon resonance. First results also came in from the CERN proton synchrotron. Reports of deuterium production at high energies had at first surprised many people but eventually all came to see that it was, in Gilberto Bernardini's words, a 'more or less banal fact'.[10] The cosmic-ray people popped up to say that they had known about it all along.

At the last session of Rochester 10, Y. Ohnuki gave a talk about 'Theories of Elementary Particles'. Some years previously Sakata had pointed out that to get isospin and strangeness all you needed as basic entities were an isospin doublet (like p and n) and a strange particle singlet (like Λ). Did that mean that those particles were the actual fundamental constituents of matter? If so, a pion would be a nucleon—antinucleon bound state, a K-meson a nucleon—antilambda bound state, and so on. Ohnuki tried to push this idea a little further. There were lots of detailed difficulties about what he had to say, but he did introduce a notion which was to have a fruitful future. Ohnuki knew about Lie groups and he suggested that the unimodular group of order 3 (essentially the group of operations shuffling three objects) would be a useful one to think about. Ohnuki had not got the right constituents but he had got the right group. It would not be too long before the correct application of it would be discovered.

The closing address at Rochester 10 was given by Heisenberg. He said, 'We could learn from the discussions that perhaps the opinions of other physicists were not quite as foolish as we thought, and we came out with the hope that the other physicists will have the same favourable impression of our own opinion.'[11] That was no bad epitaph for the last Rochester Conference to be held on home ground.

The town had been glad to see the high-energy physicists back again and it had laid on a programme of entertainments to provide relief from the conference sessions. There was also a trip to Niagara Falls. The buses used proved to be old and slow and it was something of a long, drawn-out expedition. It clearly gave some of the Russian delegates mild satisfacton to see that American technology was not always one hundred per cent successful. When their turn came around again to host the Conference, they too were to provide an equally fatiguing excursion.

11 *Geneva 11*

It had come to be felt that a fullblown international conference every year was a bit much. Accordingly a two-year gap intervened before the Conference reconvened in Geneva, on the 4−11th July 1962. CERN was again the host and this time on its own premises. Moravcsik, once more reporting for *Physics Today*, said, 'Fortunately for elementary particle physics, no new higher energy machines have been built in the last two years.' As a result, effort had been devoted to thorough consolidation and it had produced results 'both impressive and exciting, well rewarding the prodigious number of man-hours and dollars (pounds, francs, rubles, etc.) spent on them'.[1] It had indeed been a very productive period, both experimentally and theoretically, since Rochester 10.

Experimentalists had produced evidence for the existence of a great 'zoo' of resonances. In the pion system there was not only Nambu's ω putting in an appearance four years after it had been conjectured, but also a number of kindred states, ρ, η, and so on. The (3,3) and other nucleon resonances were proving to be members of a large and widespread family. One of the branches of that family involved hyperon resonances (Y*s in the notation) of strangeness -1. Similarly, the ρ and ω were found to have strange brothers, the K*s. There was still some residual hesitation about according to all resonances the status of 'particle', with the conservatives feeling that particles should correspond to bumps which were not too broad − in other words, to entities which did not decay too quickly.[2] This period of rapid discovery was made possible by advances in bubble-chamber technique, particularly the construction of liquid-hydrogen chambers and the development of computerised scanning and analysis routines.

These discoveries related to theoretical investigations in two ways, relevant respectively to internal quantum numbers (like isospin and strangeness) and to angular-momentum properties (the spins involved). The former was concerned with the search for a higher-symmetry group. The hope was to incorporate isospin and strangeness into a larger scheme, of which they would both be integrated parts, rather than their just being placed alongside each other in the somewhat arbitrary way of the earlier accounts. The idea was to find a group which not only contained operators changing a proton into a neutron (as isospin rotations did) but also those which, for instance, changed a proton into a lambda. Ohnuki had spotted the right group, SU(3), but not the right way to use it. Murray Gell-Mann and the Israeli intelligence colonel, Yuval Neeman, a mature student of Salam's, independently hit on the correct idea. Ohnuki and his collaborators had fitted the pseudoscalar mesons, the πs and Ks and the (then hypothetical) η, into an eightfold representation of SU(3). It worked beautifully. Because isospin and strangeness were part of the operations of the group, the mathematics specified exactly what isospin and strangeness assignments these particle should have. They were exactly those that they were known to possess. It was a precise and impressive fit. However, Ohnuki could not get the baryons right. Two changes were necessary to sort this out. First, it was not strangeness that was part of SU(3) but rather hypercharge, the name given to the combination $\frac{1}{2}(S + B)$ (cf. eqn [2] of Ch. 4). (Of course, for mesons $B = 0$, so that hypercharge and strangeness are the same for them.) Secondly, all the spin-$\frac{1}{2}$ baryons (p, n, Λ, Σ and Ξ) were to be put together to form another octet representation of SU(3). Once again, there was a perfect fit with the known properties. Gell-Mann called his version of the theory 'The Eight-fold Way', in punning celebration of its octet character. It is curious to note that he never published it in a physics journal. It was, and is, a fact that the frenetic character of high-energy physics means that people cannot await the leisurely process of printed publication to make their point and claim their priority. As soon as the ink is dry on the manuscript it is duplicated and mailed direct to physics departments all over the world. Literally thousands of different preprints are produced each year by the industrious labours of elementary-particle physicists. Most of these are subsequently published in the journals, although of course some fall by the wayside through proving to be wrong or irrelevant. It is very unusual for fame and recognition to be the cause of the non-appearance of a paper in archival form. Gell-Mann's Caltech Report CTSL-20 (1961) made so strong an impression, and was circulated and read so widely, that Murray just left it at that. I suspect that part of the reason was a cautious desire to be on the winning side whatever, and so not to commit oneself too far, too soon. We shall encounter further examples of this wary side to Gell-Mann's character.

By 1962 SU(3) was the front runner of the higher symmetry-schemes but not yet universally accepted as the definitive answer. In his report, Bernard d'Espagnat would only go so far as to describe it as 'the pet group of theoreticians'.[3] Today's students, taught such things in the theoretical physics nursery, may find it surprising that SU(3) took some time to catch on. Part of the difficulty lay in recognising what one might sensibly combine

Theory and experiment — Robert Oppenheimer talks to Cecil Powell outside the CERN cafeteria.

together. After all, the Ξ is 40% more massive than the proton, but SU(3) treats them as members of the same octet. As the theory developed one of its attractions was a rather successful idea (octet symmetry breaking[4]) which enabled one to go beyond the pure symmetry and give good estimates of the corrections necessary in allowing for its approximate character.

Two important developments in 1962 told in favour of SU(3). The ρ, ω, and K* resonances were seen to fit together into yet another perfect octet. The second development actually began to happen on the Conference floor at Geneva.

Experimentalists reported results indicating the existence of a doubly strange Ξ* resonance with a mass of about 1535 Mev. Earlier, a singly strange resonance $Y*_1$ had been reported with a mass of about 1380 Mev. The latter was thought to have the same spin 3/2 as our old friend the (3,3) (or Δ in more modern notation) at 1238 Mev. In the midst of a discussion following the presentation of the Ξ* results, Gell-Mann popped up to make a pregnant theoretical point. He spoke first about octet symmetry breaking, 'hard to interpret on any fundamental basis, but I hope such a justification may be forthcoming on the basis of analytic continuation of resonance states in isotopic spin and strangeness' (a most extraordinary and mysterious comment with no subsequent fulfilment). Murray recognised that if octet breaking was applied to the ten-dimensional representation of SU(3), it yielded equal-mass spacings as the strangeness changed within the decuplet. Experimentally, at that time one seemed to have to hand four Δs, three Y*s, and two Ξ*s, just as a decuplet required, and moreover their masses were more or less equally spaced with an interval of about 150 Mev. Was it not natural to hope for a tenth particle to complete the decuplet pattern and could one not make an intelligent guess what its mass might be? 'If $J = 3/2+$ is really right for these two cases [the Y* and Ξ*], then our speculation might have some value and we should look for the last particle called, say, Ω^- with $S = -3$, $I = 0$. At 1685 Mev it would be metastable and should decay by weak interactions into K$^-$ + Λ, π^- + Ξ^0, or π^0 + Ξ^-.'[5] In other words, the tenth particle would be triply strange but not heavy enough to decay into Ξ + $\bar{\text{K}}$ via strangeness-conserving strong interactions. Consequently, it would not be a resonance like its brothers in the decuplet, but a long-lived particle, decaying via weak interactions, and so it would be around for long enough to make tracks in a bubble chamber. Gell-Mann went on to make suggestions about how it might be produced.

When he had finished speaking the discussion flowed on without any immediate response to his remarkable prediction. It took a little while for the penny to drop, but when it did people realised that Gell-Mann had made what was perhaps the most astonishing and fruitful remark ever uttered in

Nobel lineup at Geneva 11 (Powell, Rabi, Heisenberg, McMillan, Segré, Lee, Yang, Hofstadter).

An impressive assembly of high energy physicists meets the press at Geneva 11.

a discussion at a physics conference. In due course the discovery of the Ω^- in 1964 set the seal on SU(3) and on Murray's claim to a Nobel Prize.

In his slightly pessimistic way, Moravcsik had characterised the state of physics in 1962 as one of crisis due to an excess of experimental information over theoretical ability to digest it. In fact, substantial developments were even then taking place to bring order out of chaos. SU(3) was one; the other was Regge theory. The latter grew out of the bold idea, conceived by Tullio Regge in 1959 as a calculational trick, of subjecting the angular-momentum variable to analytic continuation. I remember a research student of mine, Michael Fowler, telling me about Regge's work at the time that it first appeared. 'How amusing,' I remarked and changed the subject of the conversation. Like most theoretical physicists (including, I think, Regge himself) I did not realise quite how powerful a mathematical tool had been forged. It seems to have been Stanley Mandelstam alone who recognised the potentiality of the new method.

It is not very easy to convey to the reader unacquainted with complex-variable theory the substance of what is involved. He will have to trust me not to mislead him. In any question of analytic continuation it is always the singularities which are of prime significance. Remember that the only function free of them altogether is just a boring constant. The initial hope was that in angular momentum the singularities might be particularly simple; that in fact they might just be the most primitive of all singularities, the pole (cf. p. 58). Regge poles would be singularities in l (the angular momentum) at positions which depended on the energy-squared variable, s. One could write an equation for the location of a pole as s varied

$$l = \alpha(s). \tag{1}$$

Such a relation was called a Regge trajectory. Two distinct types of physical significance could be associated with a trajectory. When it coincided with a physically acceptable value of l (0, 1, 2, ...) at a positive value of s, then it corresponded to a particle pole occurring at that energy and with spin l.[6] For negative values of s (corresponding to momentum transfer in the crossed channel of energy t, cf p. 74) the trajectories were found to control the behaviour of the scattering amplitude at large t. If $\alpha_1(s)$ was the rightmost pole trajectory in the complex l-plane (the so-called leading trajectory), the large-t behaviour would be given by

$$t^{\alpha_1(s)}. \tag{2}$$

It was this crucial point that Mandelstam picked up. At last, one began to have a theoretical understanding of what controlled high-energy behaviour. Regge theory soon caught on. Gell-Mann told Geneva 11, 'My interest in the Regge pole hypothesis became intense during 1961 and has remained so.'[7] At the Conference, the interest centred on three topics.

The first concerned the high-energy behaviour of the total cross-section. I relegate some detail to Appendix 1 and here content myself with recording that the total cross-section will be controlled by the leading trajectory with the quantum numbers of the vacuum (that is, charge zero, strangeness zero, etc.). This trajectory was called the Pomeron (in tribute to Pomeranchuk). To give the constant total cross-sections which seemed to be observed, it had to pass through 1 in the forward direction,[8] namely at $s = 0$

$$\alpha_P(0) = 1. \qquad\qquad\qquad [3]$$

One can go on to discuss the behaviour of the diffraction peak, that is, the behaviour of scattering near the forward direction. If the Pomeron singularity really is a simple pole, a clear prediction emerges. The peak will shrink with increasing energy. That is a rather unexpected behaviour, since when translated into a picture of the behaviour of the target causing the scattering it corresponds

Weisskopf and Peierls ruminate. ('Vicky' was recovering from an accident).

to an increase in the size of that target with energy, accompanied by an increasing transparency, so that the total effect (cross-section) remains constant. Nevertheless, it was reported at Geneva II that just such a narrowing of the diffraction peak had been observed. This generated great enthusiasm for Regge theory. Geoff Chew told me some years later that he realised, as he sat there in CERN listening to the plaudits of the experimentalists, that the rejoicing might prove to be somewhat premature. Often in physics an idea comes along which in its first crude expression corresponds to the coarse grain of experiment, but which requires considerable refinement, both in its theoretical articulation and in its experimental confirmation, before it can lay claim to being well established. That refining process is often long and tedious and frustrating. That was to prove the case for Regge theory.

The second application of the idea was to particle poles, or rather to the appropriate generalisation to the case of resonances, taking account of their instability by displacing the poles off the real axis. In this instance the trajectories only make themselves apparent at physical, integral, values of l. They are like moles, burrowing underground and only throwing up an occasional molehill to manifest their presence on the surface. By Geneva II one could at least start to plot such points and see how they might be connected up (like children joining up dots to make a picture). An intriguing pattern began to emerge. Because there were not too many points (resonances) yet available, people tended to opt for the simplest possible curve, a straight line. What was surprising was that all the different lines, corresponding to the various possible sets of non-vacuum quantum numbers, seemed to have a common value for their slope. All came out at about one unit of angular momentum per Gev^2 of s. Later on, as more points accumulated corresponding to higher l-values, the straight line guess proved to be astonishingly accurate. Thus there seemed to be a strikingly simple structure to Regge trajectories. At the time, no one had any notion of why this should be so.

Despite all this activity there were no very good fundamental grounds for applying Regge theory to relativistic quantum theory at all. Its origin lay in the vastly different field of non-relativistic potential theory. There it did have a firm basis, but it also had something of the character of being an unfamiliar way of looking at already familiar facts. Talking about some of the Regge plots Mandelstam said they 'have been called Regge trajectories but they might well have been called Bohr trajectories'.[9] He was pointing out that the idea of families of related states with increasing angular momentum had been a feature of the earliest quantum mechanical treatment of the hydrogen atom.

1962 saw the beginnings of the attempt to ground Regge theory in quantum-field theory. Ben Lee and Ralph Sawyer studied the Bethe−Salpeter equation (the nearest relativistic analogue of the Schrödinger equation) and showed that in the simplest approximation it gave Regge poles. A group of Italians, Amati, Fubini and Stanghellini, used the ideas of peripheral dynamics (p. 91) to do the same. They went on to study more complicated cases and concluded that there might well be cuts in the complex angular-momentum plane, as well as poles. These AFS cuts (authorial abbreviations again) were mentioned in passing at Geneva 11. They had some counterintuitive properties[10] and it was not clear that they might not be cancelled out when further possible processes were considered.

From 1962 I worked for a few years on these problems. The notion was to use Feynman integrals as a sort of 'theoretical laboratory' in which to test Regge ideas in relativistic quantum theory. One could devise techniques for working out the high-energy behaviour of Feynman integrals. The results almost always involved powers of $\ln s$. (For traditional reasons I am now using s as the variable whose high-energy behaviour is to be discussed.) However, when one added up the contributions from a suitable infinite sequence of diagrams (in the first instance simple 'ladder' diagrams representing iterated interactions) these logarithms exponentiated to give a power of s, just as Regge required. Of course, all this was heuristic − that is, it was intuitively suggestive but not logically coercive. Given the intractability of a completely rigorous analysis, this use of perturbation theory as a model was a fruitful way of gaining insight. The technique easily extended to the discussion of much more complicated diagrams than just ladders. It yielded a two-line proof that the AFS cuts were indeed cancelled out. Nevertheless, there were cuts around which refused to disappear. Mandelstam had identified the likely diagrams. They were connected with an essential (equals very nasty) singularity, which Gribov and Pomeranchuk had discovered as another threat to Regge simplicity.

As with the Mandelstam representation, so with complex angular momentum, in the end everything turned out to be much more complicated and subtle than early hopes had taken into account. There are still many unresolved problems in Regge theory. Eventually, these complexities found experimental counterparts in the observed high-energy behaviour. The apparent simplicities of Geneva 11 proved to be fool's gold. A later generation of accelerators and storage rings showed that the 'medium-energy' regime of the early sixties had been delusively simple.

Meanwhile, Geoff Chew had seized the opportunity provided by these sallies into the complex angular-momentum plane to run up to the masthead a flag boldly proclaiming the rule of 'nuclear democracy'. He was now more than ever convinced that nothing was more elementary than anything else, that everything was made out of everything. The bootstrap idea drew new life from Reggeism. 'It is becoming more and more apparent that all the strongly interacting particles and resonances stand on a dynamically equivalent footing.'[11] A key new notion was summarised by the slogan 'saturating Froissart'. Marcel Froissart was an elegant French theorist who had established on very general grounds of analyticity and unitarity that total cross-sections could never increase with energy faster than $(\ln s)^2$. This was very close to the apparently observed behaviour of constant cross-sections. (Logarithmic increase is a very slow increase.) Chew, therefore, supposed that the otherwise arbitrary strength of the strong interactions must be determined to be just that which would cause the total cross-sections to bump against the bound decreed by Froissart (whose name Geoff in those days pronounced in a manner more American than Gallic). In terms of the simple Regge pole picture this required the Pomeron to pass through 1 at $t = 0$. All that was then needed was a calculational scheme to articulate these ideas. 'Unfortunately the current stone age techniques for investigating our postulates forces one to start with a definite dimensionality of the S-matrix.'[12] In other words, one had to assume how many 'things' made up 'everything'. In practice, the great Berkeley programme for fulfilling, in Feinberg's ironic words, the 'Leibnitz programme'[13] of determining the best possible strong-interacting world, was almost entirely confined to the much humbler task of trying to discuss the pion–pion system. Gell-Mann once said to me, 'That's the trouble with Berkeley. They're obsessed with pions. They suffer from piorrhea.' Even

East meets West at Geneva 11 (Blokhintsev – USSR, and Gugelot – Holland).

103

within those modest limits it proved astonishingly difficult to extract successfully so simple a consequence as the ρ resonance.

SU(3) and Regge theory were to prove absorbing subjects for several years of research. Yet a poll of those leaving Geneva would probably have established agreement with Moravcsik's judgement that 'the most significant and at the same time the most difficult experiment of the past two years has been the demonstration that the neutrinos associated with muons and the neutrinos associated with electrons are different particles.'[14] The puzzling redundancy of muon and electron was extended to neutrinos also. Nature used two where economically one might have seemed enough. The investigation of this question had been a competition between CERN and a group from Columbia University working at Brookhaven. The Americans won hands down. The difficulty of the experiment lay in the need to filter out random background events which might mask or contaminate the very rare interactions of those elusive neutrinos. After all, it was only in 1956 that Reines and Cowan had been able to exhibit the first undoubted neutrino interaction, thereby giving overt demonstration of the existence of this particle, first hypothesised by Pauli in 1929. Now people were attempting to do experiments with neutrino beams (a development with a very fruitful future ahead of it). It was essential to provide sufficient material shielding to protect the neutrinos from other particles straying into the observational area. 'The required amount of shielding, unheard of till then, was provided by the obsolete US battleship *Missouri* which was cut into iron plates which were shipped to Outer Mongolia, the name given to the area of Brookhaven where the experiment was set up.'[15] The neutrinos were produced from pion decay and so were known to be associated with muons. Their interactions were observed to see if they could, on absorption, produce either a muon or an electron. If there were to be only one type of neutrino, it should be omnicompetent. If each lepton had its own type of neutrino, then the muon neutrinos would only be capable of producing muons. After eight months of experiments twenty-nine neutrino interactions had been identified. They all gave muons. The result was clear: there are two neutrinos.

Pontecorvo had suggested that this might be so. At Geneva 11 he made another point of great significance for the future. The Brookhaven experiment had involved charged currents (a neutral neutrino had changed into a charged muon). Might there not also be neutral currents (giving an interaction in which a neutrino stayed a neutrino)? These would obviously be much more difficult to detect and at the time theorists had no prejudice in favour of their existing, but it was 'a process of great importance in its own right ... independently of theoretical prediction'.[16]

Okun, the theoretical rapporteur on weak interactions, listed a

The new periodic table (and is that Murray Gell-Mann, the new Mendeleev, on the right?).

bibliography of 285 papers which had appeared on that topic since January 1961. Despite all that weight of paper he thought that there had been 'almost no marked progress in weak interaction theory since the 1960 Rochester Conference'.[17] On the experimental side the subject was going through one of its recurrent periods of self-doubt, with hardy perennials like the $|\Delta I| = \frac{1}{2}$ rule seeming again to be in jeopardy. It proved to be a transient fluctuation of experimental fortune.

There had, as a matter of fact, been an important theoretical development in 1961. It related to spontaneous symmetry breaking and at first it appeared to be unfavourable to the prospects of that idea. My Cambridge colleague, Jeffrey Goldstone, had spotted a snag. He showed how the application of symmetry breaking could lead to the existence of massless bosons in the theory. That was a severe problem, for these Goldstone bosons would not correspond to anything actually occurring in nature. Jeffery is a very deep thinker with a marked reluctance to publish his ideas. (He once told me that he had long realised that an SU(3) octet would fit the baryons!) He was later hijacked by Salam and Steven Weinberg and forced to carry the argument further in some joint work.

The concluding address at Geneva 11 was given by Vicky Weisskopf, a tall craggy immigrant American, then Director-General of CERN. Looking over the progress of the last two years he said that previously, 'The cross sections were as flat as a desert and particle distributions as barren as phase space [what you get when nothing special happens].' In other words, it had appeared to be a featureless landscape. Now all had changed. Since 1960 'a new world has been created; you see peaks and valleys everywhere, resonances have grown all over the scale; a new world has emerged'.[18] His assessment was that, 'This Conference was an experimental conference. Most of the interesting and exciting things happened in experimental sessions.'[19] With hindsight, that appears to be only half the truth.

12 *Dubna 12*

The International Conference on High Energy Physics returned to Russia in 1964, meeting at Dubna on the banks of the Volga some 70 miles north of Moscow, on the 5th−12th of August. Dubna was a town of 'ten to twenty thousand, depending upon whom you talk to'.[1] All of those attending were housed in the same hotel, where the restaurant service was astonishingly quick by Russian standards, it being possible to complete a meal within the hour. Russian and English were again the official languages and simultaneous translation proved still to have its pitfalls. As one listened to the English on one's headphones a recurrent experience was to hear the phrase 'this depends critically upon . . .', completed by an indefinite murmur before the translator, baffled by the vital technicality, hurried on to the next sentence.

The American experimentalist Val Fitch had brought with him from Princeton a device of wider interest than the paraphernalia of high-energy physics. These were the early days of the frisbee and Fitch and some of his friends gave virtuoso evening performances on the plaza outside our hotel. They commanded the admiration of a large crowd but this was not Fitch's only impact on the Conference. He and his collaborator Jim Cronin had made a capital discovery, published just before the physicists assembled at Dubna. By means of a very skilful and delicate experimental procedure they had observed two-pion decays of the long lived-neutral K meson. According to accepted theory (p. 68) the latter was held to be *CP* odd. If *CP* were conserved, then such a decay would be forbidden. People had got so used to being consoled by the thought that *CP* provided some sort of reflection invariance that it was a nasty shock to learn of this new effect (present at the level of a fraction of a per cent, which is why Fitch and Cronin had to be so clever to detect it and so courageous to try a difficult experiment that everyone expected to give a null result). It began to be felt that nothing was sacred. The theoretical rapporteur, Sam Treiman, said of this 'most striking new development' that one had heard 'over the clinking of vodka glasses a great deal of agitated discussion'.[2] There were subsequent attempts to offer an alternative understanding of the effect. The most interesting and respectable

107

was due to John Bell and John Perring. They pointed out that from a global point of view the experiment was not conducted in a *CP* neutral setting, for the laboratory, and indeed the whole galaxy, were made of matter rather than of antimatter. If there were to be a weak long-range force whose character discriminated between particles and antiparticles, then the effect detected could be environmental in character rather than intrinsic. This Bell−Perring proposal fitted all the known facts. It also made a clear prediction. Double the energy at which the effect was observed and you would see it four times as strongly. The experimentalists obliged with the appropriate test. The effect was exactly the same. Back to the drawing board for Bell and Perring.

That little story illustrates the hazardous hostages to fortune offered by theoretical speculation. If nature had been otherwise, Bell and Perring might well have shared in a Nobel Prize. As it is, their theory is a forgotten curiosity. Yet the imagination and insight involved in making the suggestion were independent of the eventual outcome.

Fairly soon the dust settled and *CP* (equivalently *T*) non-conservation remains on the agenda. It has only been observed directly in the K-meson system, which is a very sensitive indicator for showing up the effects of small differences. The glimmering of a 'natural' explanation has had to await very recent ideas about quark families. Today, ironically, the presence of significant *CP* violation in the early universe is invoked as the explanation of why the world we inhabit is predominantly matter rather than antimatter[3] − a precise reversal of the Bell−Perring theme.

The theory of weak interactions was continuing to develop fruitfully in other directions also. Nicolo Cabibbo had clarified the issue of how to define the currents involved, by extending the idea of conserved vector currents into the domain of SU(3) and by supposing that all currents (both *V* and *A*) behave like octets under the transformations of the group. The puzzle of the weakness of the hyperon beta decays could then be solved by the recognition that it stemmed from the smallness of what has come to be called the Cabibbo angle, a measure of the balance between strangeness-conserving and strangeness-changing components which could be defined naturally in terms of the scheme. Gell-Mann further extended these ideas and thereby created the subject of current algebra. The basic notion was to specify the properties of the currents by taking certain mathematical structures (commutators) abstracted from a simple model and then to treat these algebraic relations as the foundation of the theory. In analogy with *haute cuisine* in which a pheasant is cooked with slices of veal to improve its flavour and the veal is then discarded, Gell-Mann equated the model with the veal and the commutators with the pheasant. The study of these relations, their derivation and their consequences, rapidly became a boom industry from 1964 onwards. When my former student, Bruno Renner, a man of encyclopaedic reading and conscientious thoroughness, came in 1968 to write a monograph on this topic he listed 516 papers. High-energy physics has always been prone to the bandwagon effect. Hindsight suggests that the current algebra episode was enlightening rather than fundamentally transforming.

The disposable 'veal' of Gell-Mann's culinary metaphor was the quark model which he and George Zweig had independently hit upon in early 1964. At Dubna 12 Gell-Mann said, 'The success of the 8-fold way has encouraged speculation that there exist hadrons corresponding to the spinor representation of SU(3) as well as tensor representations.'[4] What that meant was as follows: SU(3) is the group which 'shuffles' in a particular way three objects. Does that not suggest that behind its patterns there might lie a set of three basic entities? Mathematically, that is certainly the case. The three are the spinor representation to which Gell-Mann was referring. All other representations (patterns) can be formed from them by a limited set of mathematical operations. What Zweig and Gell-Mann were suggesting was that this undoubted mathematical fact might have a physical counterpart. In that case the triplet would be cast for the role of fundamental constituents. Gell-Mann called them quarks (a learned joke from *Finnegan's Wake*: 'Three quarks for Muster Mark'); Zweig called them aces, a name which did not catch on. The approaches of the two authors were somewhat different. Gell-Mann's motivation was his interest in currents; Zweig was more concerned with the constituent picture of how his aces would compose the known hadronic multiplets.[5] Each faced the same difficulty for any realistic interpretation of what they were proposing.

SU(3) fixes the electric charges of the particles associated with its representations. When one applied the rules to quarks they yielded fractional answers. In modern notation, there was a u quark of charge 2/3 and d and s quarks of charge −1/3. The s differed from the d by having hypercharge −1; it was the strange quark. u and d were of zero hypercharge and formed an isospin doublet. All known hadrons had integral charge. The baryons were to be thought of as made up of three quarks (e.g. the proton was 2 u quarks and a d quark; the Ω^- was three s quarks). Mesons were made up of a quark−antiquark pair (e.g. K^+ was $u\bar{s}$, etc.). Thus there was no problem with the known particles − in fact, quite the reverse since it followed that baryons had to be in decuplets, octets or singlets and mesons in octets or singlets, which was exactly the observed pattern. The difficulty was where were the quarks themselves? Since charge is absolutely conserved one of these fractionally charged objects must be stable, for the lightest could not decay into anything else. Yet no one had ever seen a fractionally charged particle. Hunting the quark soon became a popular pastime. I have written elsewhere:

A feverish search began to find them, inspired no doubt partly by the thought that the first man who indubitably discovered one might find that he had also earned himself an enjoyable trip to Stockholm. The technique for the search was essentially based on highly ingenious refinements of the idea used by Millikan in 1910 to measure the charge on the electron. The difference, of course, was that this time it was hoped that fractional answers would be found occasionally, corresponding to the presence of quarks. It was thought that cosmic rays would have been showering the earth with quarks for many aeons and people tried to work out where they might tend to accumulate in quark-rich deposits. The bottom of the ocean was a good bet and quantities of mud were sucked up to be submitted to analysis. Others reckoned that the earth's magnetic field would help to focus this rain of quarks into the polar regions, so that chunks of arctic ice would be worth investigating.[6]

Despite occasional excitements, it is sad to relate that no bona fide claim to quark discovery has yet been substantiated. It will emerge in due course what the theorists make of that.

In the meantime, Gell-Mann was characteristically cautious. For many years it was his habit to refer to the 'presumably mathematical' quark. I always considered that to be a coded message. It seemed to say, 'If quarks are not found, remember I never said they would be; if they are found, remember I thought of them first.' His caution was matched by a certain reserve in the theoretical community about the whole idea. Zweig commented later that the reaction to the proposal was 'generally not benign . . . . The idea that hadrons were made of elementary particles with fractional charges did seem a bit rich.'[7]

At Dubna 12 Salam gave the rapporteur talk on 'Symmetry of Strong Interactions'. He exhibited a slide listing all the contributors to the session and remarked that the Russian language had a useful phrase 'bratskaya mogila' — 'the friendly communal grave'.[8] Not all the ideas he referred to were to be condemned to mouldering oblivion. The proposal of a fourth quark, carrying the quantum number facetiously called charm, was beginning to be made, either in an attempt to get around fractional charge (not an idea with a future) or to establish a correlation between two pairs of leptons (electronic, muonic) and two pairs of quarks (u, d; c, s), which would prove to be an eminently resurrectible concept in due course.

July 1964 had seen a flurry of activity related to yet another way of exploring group theory in elementary particle physics. SU(3) is not the only group of special unimodular type with which one might be concerned. Rotations in three dimensions have the same mathematical structure (are isomorphic to, as the mathematicians say) the smaller group SU(2). The theory of angular momentum precisely deals with behaviour under these rotations.

110

If one puts the two groups together one gets what is called a direct product, SU(3) × SU(2). It occurred to the Turk, Feza Gursey, and the Italian Count, Luigi Radicati, both in Brookhaven for the summer,[9] that one might consider the possibility of invariance under the larger group SU(6). There is more to SU(6) than there is to SU(3) × SU(2),[10] and so this requirement would impose further restrictions on what could be the case. A similar idea had been tried before in the 1930s. In nuclear physics it proved to be a useful approximation in certain circumstances to treat the nuclear force as being invariant under an SU(4), obtained by conflating the SU(2) of rotations with the SU(2) of isospin. It is astonishing how often the high-energy theorists, who tend to think of themselves in a rather lordly way as pioneers operating at the frontiers of knowledge, have had to borrow ideas from 'humbler' subjects, such as the investigation of nuclear structure.

SU(6) scored an immediate success. Because it combines spin with internal SU(3) symmetry, it specifies the way the two are linked together in its representations (patterns). There is a 56-dimensional representation of SU(6), corresponding to three quarks and composed of a decuplet of particles of spin 3/2 and an octet of particles of spin ½. That exactly fits the known structure of the lowest baryons. There was similar success in fitting the mesons with a 35-dimensional quark–antiquark representation. It was a most impressive achievement. It also posed a problem which was not to be solved for some time. In the 56 the three quarks are treated completely symmetrically with respect to spin and SU(3). But they are spin-½ entities and so, by the spin and statistics theorem, their overall wavefunction must be antisymmetrical. The only way to satisfy that would be if the spatial part of that wavefunction (the only part not specified by SU(6)) were to be antisymmetrical, a result contrary to reasonable expectation for the lowest energy state, whose spatial wavefunction is invariably found to be symmetric. For the time being the problem was just noted, but a year later Han and Nambu introduced the idea of 'colour'. They proposed that quarks came in three varieties (red, white and blue was the popular joke terminology, though the primary colours, red, yellow and blue, would have been more logical) in addition to their 'flavours' (that is SU(3) labels), but that all the observed hadrons were 'colour singlets' (that is, they were colourless or 'white'). It would then be these extra colour degrees of freedom which would produce the required antisymmetry in the 56. With great prescience, Han and Nambu went on to speak of an SU(3)$_c$ of colour and of the eight vector gauge fields which would be associated with it. Except for an overingenious trick to try to use colour also to yield integral values for charges (with fractions only resulting from averaging over colour) they had the basic notion of what was to turn out to be the fundamental theory of strong interactions, quantum chromodynamics. At the time their paper made little impact, appearing to be extravagantly speculative in its multiplication of types of quark.

In the middle sixties attention was directed to a different problem. SU(6) appeared to be a lucky guess, a heuristic device which brought order but which could not itself be a fundamental theory. It was often called static SU(6), since it dealt with the properties of particles as they might be found just lying around. A dynamical theory, capable of coping with their interactions, would require a considerable generalisation. In particular, it would have to incorporate special relativity. In the latter theory, the spatial rotations of SU(2) have to have added to them the 'boosts' which take one from a frame of reference at rest to one which is moving with uniform velocity. These boosts can be thought of in a four-dimensional way as pseudorotations, mixing time and a spatial coordinate in the way that true rotations mix the two spatial coordinates at right angles to the axis of rotation. Because time really is different from space (as different as history is from geography) incorporating these pseudorotations changes the character of what is going on. A natural generalisation of SU(6) is a group called S$\tilde{\text{U}}$(12) (pronounced 'S U twiddle twelve'). Salam got very excited about it. However, it soon became apparent that there were grave difficulties. The effect of the 'twiddle' is to produce what the mathematicians call a non-compact group. (A compact group 'moves' one round a finite surface, like the surface of a sphere; a non-compact group 'moves' one around an infinite surface, like the surface of an hyperboloid.) That turns out to result in fatal consequences, such as ghosts, which destroy the possibility of consistent physical interpretation.[11]

A year after Dubna 12 there was an extended summer seminar at Trieste, in the successful International Institute which Salam had founded and directed, principally to help physicists from developing countries to keep in touch with modern ideas. Gell-Mann came and gave a talk about these symmetry questions. Efforts to salvage S$\tilde{\text{U}}$(12) were still going on but Gell-Mann made no reference to the group, despite the presence of Salam sitting in the front row. Most of us recognised the implicit judgement thereby conveyed. An ingenuous young Englishman, who worked in the more abstract reaches of quantum field theory, felt aggrieved on behalf of his professor and asked Gell-Mann why he had not mentioned S$\tilde{\text{U}}$(12). Murray paused for a moment and then said, 'Oh, you mean the twiddle-twaddle.'

There is a footnote to all this. One desperate remedy which would have saved S$\tilde{\text{U}}$(12) would have been to extend the dimensions of spacetime from 4 to 12. Although an analogous proposal had been made by Kaluza and Klein in the 1920s, it was not an idea that raised any enthusiasm in the 1960s. In the more boldly speculative, not to say rash, 1980s, the enhanced dimensionality of spacetime has made a comeback. The unobserved dimensions are supposed to be 'compactified', that is to say, rolled up out of sight, rather like a thin straw which, though two-dimensional, looks linear because its cross-sectional diameter is small.

Regge theory rumbled on. V.N. Gribov, on whom the mantle of Landau had descended after the latter's sad incapacity following a road accident, described a new method of calculation, the Reggeon calculus, which would become the standard technique for wrestling with the complexities of cuts. It had been thought initially that if a particle lay on a Regge trajectory that was a sign that it was a composite object and not a fundamental entity. Gell-Mann and Goldberger suggested that this was not necessarily so, for it seemed that vector mesons could cause elementary particles to be swept on to Regge trajectories. I contributed to the Conference an involved calculation which supported that contention. I now blush to recall that piece of work. In the course of it I made two compensating mistakes. The result was correct; the reasons given were wrong.

The issues which now catch the eye on surveying the proceedings of Dubna 12 are, as always, a minority of the topics discussed. A great many of the matters fiercely debated at the time have proved to have been only of transient concern. For example, there was a heavy input from Russian theorists about what they called the 'quasipotential' approach to strong interactions, a retrograde attempt to reintroduce a potential-type notion via dispersion relations. Mandelstam spoke of the 'intense debate on the theme: which of the non-existent theories is better — the S-matrix theory, the usual field theory or the axiomatic approach'.[12] None of these theories in the forms then considered is the basis of current theoretical thinking. In other reports one detects the signs of great things yet to come. Presenting neutrino experiments from CERN, Gilberto Bernardini said 'If something was really good about the experiment it was the beam.'[13] The enriched neutrino beam had been made possible by a clever device, van der Meer's horn, invented by a Dutchman at CERN. Simon van der Meer's sustained ingenuity was to prove the eventual basis of a spectacularly successful programme of neutrino physics at Geneva. He went on to gain a share in a Nobel Prize for further felicitous instrumental innovation.

The CERN people reported their conclusion that if there were an intermediate vector boson, then its mass must exceed 1.8 Gev, for otherwise their experiment would have revealed its presence. (We now know the relevant mass is about 80 Gev.) There is an embarrassing story behind that announcement. A year earlier, at a European Conference held in Siena, a regular feature of the scene had been the sight of the CERN physicists sitting each evening in a café, in anxious conclave about what their results actually showed. They thought they had a signal which might indicate the presence of an IVB. The dilemma was whether to go for bust and claim this important discovery or to stick with caution and await further investigation. In the end they decided to stake all on a bid for glory. Shortly afterwards someone noticed that their big detector had a flaw in its construction. It was this which had caused the anomaly interpreted as an IVB. It is easy to scoff. Experimentalists are under great pressure, both to establish priority[14] and to justify the large expenditure that is needed for their enterprises. The conference circuit intensifies that pressure by creating arenas in which announcements can have maximal impact. The temptation to avail oneself of the opportunity thus provided is hard to resist.

Off stage from Dubna 12 an important discovery had been made in 1964 which was eventually to be of the greatest significance for weak-interaction theory. Peter Higgs was spending some time in North Carolina, on leave from the University of Edinburgh. Higgs was a competent theorist, but of no great previous distinction, who was now to hit on an idea of the highest importance. In the words which Oppenheimer used when he learnt of Higgs's theory, he was to 'lay the ghost of the Goldstone boson'. The latter, you recall, was the disastrous zero-mass particle which appeared to threaten a physically realistic use of the idea of spontaneous symmetry breaking in elementary particle physics. Once again, another branch of physics had something to teach the high-energy theorists. Spontaneous symmetry breaking plays an important role in solid-state physics without there being any sign of Goldstonian troubles. Phil Anderson drew attention to this and suggested that this pleasing state of affairs was due to the presence of the long-range force of electromagnetism. Higgs then found a model which displayed the essential features relevant to elementary particle physics. He took a theory with four field components in it: two scalar fields and the two (transverse) polarisation states of a massless vector field (the analogue of the electromagnetic field). The application of spontaneous symmetry breaking produced a reinterpretation of the four-component structure: a massive scalar field and a three component massive vector field (since vector particles with mass possess both transverse and longitudinal polarisation states). In other words, in an astonishing but highly satisfactory manner, the threatened massless Goldstone boson and the massless vector field had gobbled each other up to produce the states necessary to describe a massive vector field. It was an extraordinarily beautiful resolution of the problem, soon extended by Tom Kibble to the case of (non-abelian) Yang–Mills fields. Perhaps one of the most surprising aspects of the story is that this idea did not occur to the acute and fertile mind of Jeffrey Goldstone himself.

Higgs's mechanism makes essential use of scalar fields, which were dubbed 'Higgs bosons'. There have been speculations about whether Higgs bosons might be no more than ciphers for a deeper underlying dynamical mechanism (there are some analogies with the theory of superconductivity to encourage this view) but no one has succeeded in showing how this might be the case. The conventional expectation, therefore, is that there are these Higgses (as they say) to add to the portfolio of fundamental particles. Their discovery would put the cap on the whole idea. Unfortunately, the theory does not predict what their mass might be, so experimentalists can only keep their eyes open, without being exactly sure where to look. So far they have not seen any Higgses.

There had been a running joke at recent conferences on particle physics. It started in 1962 at Geneva 11 with the reproduction of a *New Yorker* cartoon. Two archaeologists are looking at a tiny pyramid poking out of the sand. One says to the other 'This could be the discovery of the century — depending, of course, on how far down it goes'. The application to high-energy physics needs no elaboration. In 1963, at a conference at Stanford at which the demise of a simple Regge pole-only picture had been clearly recognised, the cartoon made a reappearance with a new caption. This time it read: 'If this is what I think it is, let's cover it up and forget it.' At Dubna 12 Salam made what proved to be the final contribution to the pyramid-joke series. His slide showed a large pyramid balanced on its apex. The caption read: 'I hope this structure holds till the next conference.' High-energy physicists have always been a cheerful bunch of people.

That good humour was put to the test at Dubna 12 during the Conference outing. Nick Samios discreetly refers to 'a memorable boat trip to Kalinin'.[15] It was certainly an unforgettable experience. We set off with the intention of having a picnic at a bathing station on the Volga. Hours went by and no such *plage* hove in sight. It was afternoon before we got there. It was clear that we would take quite as long to return and consequently the boat turned round immediately, pausing only long enough to allow leading Russians, like Bogolyubov and Blokhintsev, to leave us and embark in a fast motorboat for their return journey. It was one of the principal consolations of the long way back to pass that motorboat, broken down on the way. Long before we disembarked at Dubna, all food on board had been consumed. Shared adversity is a great source of comradeship. The British theorist Paul Matthews, a cheerful companion in any circumstance, led a session of community singing. Once ashore we asked what had gone wrong. It was rumoured that the captain had not been there before and it had proved to be further than he had expected but Maurice Jacob tells me that the organising committee had mistaken the one-way time for that of the round trip. I must say, however, that a splendid meal was awaiting us in the hotel as a consolation for the privations of the day.

13 *Berkeley 13*

The struggle to adapt the format of the Conference to the expanding nature of the world of high-energy physics was taken a stage further when the 13th Conference convened at Berkeley, California, for the period 31st August—7th September 1966. Participants were permitted to submit papers (and the 500 of them produced an average of one apiece), but there were no organised parallel sessions in which they were to be presented. Instead, two and a half days of informal discussions were followed by three days of formal rapporteuring, with the intervention in between of a necessary break for digestion and critical assessment. This scheme was to prove the beginning of a new pattern for the future. Another innovation tried at Berkeley did not turn out to have equal staying power. In an apparent attempt to violate causality, Murray Gell-Mann was invited to give the Conference summary at the beginning. He started his keynote address by saying:

> Some people have said 'I hope you can tell us everything that's going to be important so that we don't have to go to the sessions' ... Other people have told me that I'm far too young and far too involved with the subject to be able to give any general philosophical pronouncements and that I should concentrate on some discussion of what's going on in the field. I feel the last is probably the most reasonable.[1]

Looking back at the first Rochester Conference, he said that then people were involved in an 'unreliable calculation of deuteron structure based on the exchange of one pion'; now they were involved in 'unreliable calculations of 50 to 60 bound states on the basis of exchanging 50 to 60 particles, and the progress is amazing'.[2] Gell-Mann said that the experience of attending the conference was not one of instant illumination of the scene but that 'going home afterwards one has a lot of ideas that have seeped in during the meeting and that form a coherent picture in the mind'.[3] I think that is right, and no doubt even more so for those of us not so quick on the uptake as Murray. Attending the 'Rochester' Conferences leavened the lump of one's everyday

116

thinking and started a ferment in the mind which continued in the months following.

Gell-Mann exhorted his audience not to waste effort in arguing S-matrix theory versus field theory but to concentrate on relativistic quantum mechanics and to devote their energies to seeking the construction of further high-energy accelerators so that 'we can really learn about the basic structure of matter'.[4] Reggeism was undergoing a revival, despite the complications of cuts, but processes at large momentum transfer and at large multiplicities were proving hard to analyse. (It was precisely in these regimes that important insights were later to be found.) Gell-Mann went on to speak of 'those hypothetical and probably fictitious spin-½ quarks'.[5] Attempts to talk about a realistic quark model of hadrons had, in those days, to decide whether the quarks were heavy, and so energetically expensive to produce, or light and contained within a deep potential barrier which hindered their individual release. Anyway, a realistic quark model was hard to believe about hadrons 'since we know that, in the sense of dispersion theory, they are mostly, if not entirely, made up of each other'.[6] Gell-Mann was always careful not to write off the bootstrap completely. His conclusion was that 'whether or not real quarks exist, the q and $\bar{q}$ we have been talking about are mathematical; in particular I would guess that they are mathematical entities that arise when we construct representations of current algebra. ... one may think of mathematical quarks as the limit of real light quarks as the barrier goes to an infinitely high one.'[7] The last phrase proved prophetic.

Things were getting sorted out in relation to CP violation, 'the most exciting topic ... that was opened two years ago'.[8] An attractive notion was the superweak theory of Lincoln Wolfenstein. It envisaged an interaction twice as weak as conventional weak decays and changing strangeness by two units. Only in the delicate circumstances of the neutral K-meson system would its effects be amplified sufficiently for them to be manifested at an observable level. (That resulted from the enhancing effect of the very tiny mass difference between K_1^0 and K_2^0.) Thus Wolfenstein predicted that the only *CP*-violating effects to be found experimentally were those which had already been discovered. At Berkeley 13, Fitch was able to report the only new experimental information reliably available on *CP* non-conservation. The phase of the relevant K-meson amplitude had been determined, in addition to its magnitude, which was already known at Dubna 12. That was all that two years of effort had yielded (or more years were to yield) and, 'Two years is, after all, a substantial part of a physicist's professional life – an even larger fraction of his productive life.'[9]

Superweak is a phenomenologically successful idea. A completely contrasting idea had been put forward by T.D. Lee and others. They suggested

Murray Gell-Mann — a hero of our tale.

that there was a 0.1% violation of charge-conjugation invariance in strong interactions. Combined with the maximal parity-violating effects in weak decays this would produce the observed *CP* effect. Of course, the *C* violations should be observable directly and charge asymmetry in η decay was a potentially sensitive probe. By Berkeley 13 it was clear that this proposal was not in very good shape. Lee was fighting something of a rearguard action. In his own talk he complained that there were too many kinds of *CP*-violating theories but also pleaded for an open mind. 'The whole situation reminds one of cosmology; we have only one experiment but many theories. I regard this situation as unhealthy.'[10] Certainly, weak interactions continued to puzzle. Cabibbo showed a slide with two ostriches, their heads buried in the sand while a great explosion was taking place in the background labelled '*CP* violation'. The caption read, 'We fully understand weak interactions.'

Gell-Mann had said about the *C*-violating proposal, 'you can look for the new force any place; you can see it behind any shutter', and he went on to say about the experimental search, 'I don't know if it will reach the height of comedy achieved at Siena two years ago [the IVB fiasco; see p. 113], but it may be pretty good.'[113]

The atmosphere in particle physics in previous years had become extraordinarily open to speculation, following the enlarged horizon of possibility revealed by the discoveries of the 1950s. Gell-Mann told the conference that he was really looking forward to the proposal of a new hypothetical particle that was both a quark and an intermediate boson, that was charged but invariant under charge conjugation, that obeyed parastatistics[12] and was also a magnetic monopole. 'For this particle I suggest the name "chimeron", and I look forward not only to hearing it proposed but also hearing its existence partially confirmed.'[13]

Gell-Mann had seen current algebra as the way into quark theory. The subject was summarised at Berkeley 13 by Roger Dashen. It had all become pretty technical. The most recent development had been the exploitation of sum rules. These were relations resulting from taking the matrix elements of current commutators (see Appendix 2 for a little detail). Because the commutators were evaluated at equal times, and the definition of that depended upon the choice of a particular time axis, the results were not manifestly covariant in form. That is to say, they depended upon the way in which one chose the components of energy and momentum in the particle states selected. A welter of different equations resulted, none of them taking a form that seemed very useful for actual evaluation and application. An important advance was made when two Italian theorists, Fubini and Furlan, recognised that if one chose a frame of reference in which the initial and final particles were given infinite momentum,[14] wholly sensible and useful equations resulted. That was because this proved to be the frame of reference in which the momentum q carried by the currents corresponded to a fixed value of q^2, whichever terms in the sum were being evaluated. One could carry the argument a step further and choose a value of q^2 corresponding to a particle pole in the current amplitude. The residue at that pole would consist of a purely hadronic amplitude and so one obtained a sum rule for such hadronic amplitudes. Something odd was going on: from current algebra one seemed to discover a result in purely strong interaction physics, which had nothing to do with currents!

At Berkeley 13 this all seemed pretty peculiar. Later it fell into place. Lurking behind it all was Regge theory. The derivation of sum rules by the odd appeal to infinite momentum was perceived to be a roundabout way of recognising that current algebra required the existence of certain fixed pole singularities in complex angular momentum for current amplitudes, over and above the ordinary Regge poles that they shared with hadronic amplitudes. The strong-interaction results were identified as 'superconvergence relations', expressing the consequence which followed from Regge theory that these particular amplitudes decreased very rapidly at infinity. The condition that this happened proved to be identical to that which was necessary to make sense of the infinite- momentum limit, assumed in the earlier derivation. Later these superconvergence relations were themselves to lead to another significant development in theoretical exploration. All this may strike the general reader as somewhat arcane. It may be thought of as an illustration of the way in which relativistic quantum mechanics can tie together topics as apparently diverse as current algebra and Regge theory.

In an extended discussion of theoretical ideas, Francis Low asked, 'Can one formulate a clean criterion for believability of a dynamical model? Probably not. It is a question of individual judgement, of what in law might be called the judgement of the man of ordinary prudence.'[15] One thinks of that tacit skill, knowing more than it can tell, which Polanyi correctly identified as lying at the root of scientific method.[16] Nowhere was that skill more needed at the time than in the attempt to establish the resonance spectrum of hadronic physics. Experiments were now being made at high energies where many partial waves (that is, many angular-momentum states) were involved. No longer was it sufficient to look for a big bump, as it had been in the good old Chicago days of the (3,3) resonance. Individual peaks got lost in the welter of what was going on. Only a refined partial-wave analysis, seeking phase shifts passing through $90°$, could show up the existence of such sunken treasure. The exercise involved masses of data, extensive computer programmes, uncertain assumptions about how to lump unknown detail into guesses at background behaviour. Large teams grew up around the principal protagonists and it was scarcely surprising that less than universal agreement was to be found in the conclusions of the rival groups. Gerson Goldhaber introduced a session on boson resonances by saying, 'I understand that it was Fermi who once said, "with one event one can obtain a cross-section, with two events one can obtain an angular distribution, and with three events one can obtain a polarisation". What is becoming obvious at this Conference, however, is that with four events one *cannot* obtain a resonance.'[17] Those who presented work in this area seemed quite as keen to denounce the conclusions of others as to announce results of their own. A colourful figure on this scene was Claud Lovelace. Austere in appearance, booming in voice, confident in attitude, he had been notorious while holding an appointment at CERN for his habit of sitting in the front row at other people's seminars reading the pile of preprints he had brought with him, but occasionally looking up to pose a pointed question. It is said that once, when Lovelace himself gave a seminar and the rest of Theory Division by arrangement turned up, each with his own pile of reading matter, he was not amused. Claud had gone into the resonance-hunting business in a big way. There was by now so much data to fit, and so many assumptions necessary to make the task tractable, that one could always indulge in a straight-faced application of statistics to 'prove' the total unreasonableness of a colleague's view, judged from one's own perspective. Lovelace pronounced the 'best-fit' results of his principal rival, Gordon Moorhouse from Glasgow, to have only a chance of one in 10^{166} of being right. In reply, the milder Moorhouse conceded a chance of one in 10^{24} of Lovelace's correctness.

Of course, both assessments were ridiculous. Eventually the great phase-shift industry, through sustained effort over many years, attained an acceptable degree of convergence in its conclusions. The results boil down to the entries in the 'Rosenfeld' data tables, so named after Art Rosenfeld who masterminded for many years the particle data group at Berkeley, acting as both conscience and communicator for the high-energy physics community in these respects. The slim booklets so produced are a vital linkline between theory and experiment.

14 *Vienna 14*

Vienna is the capital of a great empire that has faded away. 'Ichabod' seems written over its palaces and grand equestrian statues. The Fourteenth International Conference on High Energy Physics met there, from the 28th of August to the 5th of September 1968, in the marbled vastness of the Hofburg Palace. What the location afforded in grandeur it lacked in convenience. Marble is a material better adapted to making the impression of magnificence than to providing an acoustical setting in which one can actually hear what is being said. The conference reception was held at the Schönbrunn Palace. As we queued up the grand staircase in order to be received at its top, the news spread that it would be the President of Austria whose hand we were to shake. I turned to my friend Bruno Renner, waiting beside me, and said that it would be the first time I had been greeted by a Head of State. He replied, 'Remember this is Austria. He has nothing else to do.'

In terms of physics it was a deceptive conference. What hindsight would reckon to be many of the most significant developments were either not reported at Vienna 14 or else attracted little interest there. At the Conference itself a lot of attention focussed on a new development in Regge theory called duality. The classic role of a Regge pole was to determine high-energy behaviour in one channel (say, at large s) and the location of resonance poles in the crossed channel (finite t). It was natural to suppose that a model for behaviour in the two channels taken together would involve two distinct sets of Regge poles: one for large s/finite t, the other for large t/finite s. The relevant amplitude was, therefore, expected to be a sum of two distinct terms. Studies based on finite energy sum rules (p. 120), together with other considerations,[1] suggested that this was not the case, but that there was just a single term capable of playing the dual role of catering completely for both s and t channels. That implied that the high-energy behaviour in a channel could be thought of as resulting from the successive contributions of resonances in that channel. Haim Harari was moved to ask, 'Can we describe the whole world of strong-interaction processes in terms of resonances only?'[2] In the course of 1968 Gabriele Veneziano produced a simple model which had exactly this property (see Appendix 3). An enormous industry developed in subsequent years, studying the properties of theories incorporating the property of duality. From the insight of Jeffrey Goldstone and others it came to be realised that underlying the complexities of the discussion was a simple physical picture. A lot of difficulty was encountered in making dual theories completely consistent. They were liable to be plagued with ghosts and tachyons (faster-than-light particles). The theories which really worked were identified with the relativistic quantum mechanics of strings — that is, extended one-dimensional entities, tracing out a two-dimensional sheet in spacetime — as opposed to the conventional pointlike entities, previously considered, which trace out a one-dimensional worldline in spacetime. Eventually, the phenomenological pursuit of duality, conceived as giving a detailed account of hadronic physics, collapsed under its own weight of multiplying complication, like so many of its predecessors in that ambitious task. In the mid 1970s the disillusioned largely abandoned the struggle, but the insight into the possibility of a richer form of relativistic quantum mechanics remained as a permanent gain. It proved eventually to be capable of unexpected application. There had been embarrassment that string theory persisted in producing spin-2 particles of zero mass. In the 1980s the bold move was made to turn this vice into the virtue of the graviton (just such a particle), thereby affording a promising way of attempting to reconcile gravity and quantum theory. At the same time this changed the natural length scale of the string from 10^{-13} cm to the Planck length scale of 10^{-33} cm, effectively pointlike at accessible energies. That understanding is the basis for the contemporary theory of superstrings, hailed by many today as a potential Theory of Everything. (Will theorists never learn?) One does not have to acquiesce in such grandiose claims in order to acknowledge, nevertheless, the value of exploring as thoroughly as possible the theoretical options open to enquiry.

Meanwhile, the juggernaut of experimental measurement trundled on its way. Rosenfeld (of the particle data tables) was quoted at Vienna 14 as having observed that at the present rate of measurement of two million events per year 'we can expect to see several 4σ and hundreds of 3σ fluctuations/year'.[3] In plainer words, if you shuffle and deal the pack a great many times you will from time to time deal yourself rare combinations of cards which may appear to be more significant than they should. The experimentalist who made the quotation (B. French from CERN) went on to draw the moral that what was needed was not so much more experiments but rather better ones. In his talk he referred in passing to the quark model, to which he accorded the lowly status of 'a convenient mnemonic'.[4]

There was a certain timeliness about such cautions. A great argument

'*Peef*' *Panofsky.*

was going on about the character of a resonance called the A_2. One group discerned its nature as being a split peak; others could find no sign of this double hump but saw it only as a simple bump. Theoretically, the question was intriguing, since a twin peak would correspond to a resonance which was a dipole, the square of the normal pole expectation. Lots of ingenuity was exercised in the attempt to explain this unexpected phenomenon. All in vain! The effect eventually went away, but not without rumours about how those of the dipole party were alleged to have treated the accumulating data so as to support their view. Such incidents are fortunately rare in physics.

Preliminary data were reported from Serphukov, where the Russians had constructed the highest-energy synchrotron, operating (but not all that successfully) at 70 Gev. At somewhat lower energies people were beginning to look at processes involving large transfers of momentum. The cross-sections were small, due to rapid decrease on moving away from the forward-scattering peak, but it was noted that there were signs of a flattening out at about 3 $(Gev/c)^2$. This was later to prove a very important area for investigation and analysis.

On the theoretical side the current algebra boom was at its height. Steven Weinberg reported that, 'During these two years [since 1966] I have received preprints of work on current algebra weighing altogether about 55 kgm.'[5] He went on to express 'extreme optimism'[6] about the project.

Serphukov's synchrotron was not the only new accelerator to have come on the air recently. At Stanford in California, a two-mile long linear accelerator had started to produce electrons of very high-energy. The shape of this machine was dictated by the fact that fast electrons lose a lot of energy if you try to bend them round corners (technically, through synchrotron radiation). 'Peef' Panofsky, the short, dynamic director of the Stanford Linear Accelerator Center (SLAC), gave an account at Vienna 14 of preliminary results on high-energy electron scattering off protons. In particular, he reported on what came to be called 'inclusive' inelastic scattering. This related to events in which the electrons produced new particles but, instead of looking at the details of the subsequent final states, one just lumped together all the possible things that had been going on. Inclusive cross-sections were of particular interest because of a suggestion made by a Stanford theorist, J.D. Bjorken, universally known as 'BJ'. By a complicated argument, apparently invoking current algebra and dispersion theory, Bjorken had concluded that in the limit of high-energy (ν) and large momentum transfer (q^2), the structure functions (which were the quantities determining the inclusive cross-sections) should scale, that is they should become functions, not of ν and q^2 separately, but simply of the dimensionless ratio ν/q^2. Panofsky reported that this scaling behaviour seemed actually to occur in the 'deep inelastic' regime of large

ν and q^2. He commented that it was 'indicative of point-like interactions being involved'.[7] My recollection is that this caused little stir at Vienna 14. Certainly, Panofsky's talk does not go to town on the issue. However, scaling was a result of the highest importance.

The historically minded might have cast their minds back to 1911. In that year Rutherford's group at Manchester was studying the scattering of α particles by gold atoms. It was found that some of the αs bounced back. Rutherford later said this was as surprising as if a 15″ naval shell had recoiled on impact with a sheet of tissue paper. His interpretation of this 'deep' scattering (that is, scattering involving a large transfer of momentum to the α particle, which had its direction of motion reversed) was that it was caused by a point-like concentration of positive charge at the centre of the atom. The nucleus had been discovered. In an exactly analogous manner, the SLAC discovery of deep inelastic scaling indicated the existence of point-like structures within the proton. In due time, the accumulation of data from different types of scattering, involving neutrinos as well as electrons, was to point to these point-like constituents as possessing the properties (quantum numbers) associated with quarks. In my view, the publication of the scaling data in 1969 makes that year a watershed for the quark model of the structure of matter. Thereafter, with increasing compulsion, quarks just had to be taken seriously as a realistic account of what was happening inside hadrons. Yet lurking behind the discussion was still a theoretical embarrassment. If the electrons at SLAC really did bounce off the quarks, why did they not expel them from the proton in the process? Until one either found individual quarks or some way of confining them, there was bound to be lingering uneasiness.

The simplest way to think about scaling results was introduced by Feynman during a visit to SLAC shortly after the Vienna conference. He proposed a model of constituents which he suggested should be called by the neutral but distasteful neologism, 'partons'. Feynman's parton model was formulated with great intuitive skill and not a little oversimplification. It used a picture borrowed from nuclear physics, called the impulse approximation. This supposed that in the deep inelastic regime the partons could be treated as if they were freely moving particles, unrestrained by the forces that held the hadrons together. By a simple calculation this led to the scaling result.

Bjorken's analysis was pioneering but murky. Feynman's model was transparent but relied a little too much on his inimitable hand-waving. Eventually, the property of scaling called for the recognition of the behaviour called asymptotic freedom in quantum field theory (see later, p. 142). That analysis also predicted that scaling was not quite exact (as indeed it was found not to be) and that some subtle residual logarithmic variation with energy would also be a feature of the phenomenon. That correction, however, was of the nature of a refinement rather than a refutation, and the quark model remained as a continuing insight. In the interim (early 1970s) I think it would be true to say that the most satisfactory account of how to think about point-like hadronic constituents was provided by the covariant parton model developed by a group of us in Cambridge. To say that is not to be seized by *folie de grandeur*, for the claim is, in fact, a pretty modest one. The essential ideas were Bjorken's and Feynman's; all we did was to use a bit of expertise to find the tidiest way to think about them. It was for me an interesting change of activity, for I had spent the second half of the sixties mainly collaborating in quite elaborate attempts to understand the singularity structure of S-matrix theory and of Regge theory. There was an earthiness about parton modelling, greatly encouraged by the phenomenological concerns of Peter Landshoff with whom I was collaborating, which was refreshing after the abstract perplexities of the complex plane. Yet the latter activity had revealed a beautiful structure which is part of the fabric of relativistic quantum mechanics − although it also proved to be a structure too subtle and complicated to be readily cashable in terms of experimental implications. I sometimes hope that one day some clever person will find out what to do with it and that we primitive explorers of singularity properties will then be accorded a modest degree of posthumous praise.

Moravcsik wrote up Vienna 14 for *Physics Today*, the last time that high-energy physics was reviewed in that particular way. He wrote, 'Not many years ago the theoretical trends in high-energy physics were concentrated on efforts that had very little to do with experiments.' Moravcsik felt that things had improved in that respect and he expressed his pleasure that this was so. 'It is impossible to overestimate the importance of this relationship for the morale of experimentalists, and for keeping theoretical physics from becoming only a mathematical exercise.'[8] Ironically, one of the most important developments linking theory and experiment went totally unreported, not only at Vienna 14 but also at its successor Conference in 1970.

The new development lay in the construction of a unified theory to account for both weak and electromagnetic interactions, a synthesis as remarkable as the nineteenth century's unification of electricity and magnetism. People had been attempting it on and off for a while, but it was a tricky enterprise. Weak interactions must be allowed to violate parity whilst electromagnetic interactions must be scrupulous to respect it. The great disparity of apparent strengths required the intermediate vector bosons mediating the weak interaction to be extremely heavy, in comparison with the masslessness of the photon. Spontaneous symmetry breaking provided the opportunity to cope with the latter problem. This was the ingredient which had been missing in earlier attempts. In 1967 Steven Weinberg published what

Abdus Salam making his point.

was to become the 'standard model' for electroweak unification. The same idea had occurred independently to Abdus Salam, but his first public announcement was in the rather odd setting of the Nobel Symposium held in Sweden in May 1968. The Salam—Weinberg theory involved the group SU(2) × U(1). Its least appealing feature was the need to fuse two groups in a direct product instead of just being able to employ one. This resulted in there being a phenomenological parameter, essentially determining the balance between the two groups, which has to be read off from nature. This parameter is expressed in angular form and is usually called θ_w, the weak or Weinberg angle, according to taste. Eventually it was found experimentally to have the value 28°. In due course, Salam and Weinberg were rightly given shares in a Nobel Prize for their discovery.

Sidney Coleman has said of this theory that 'rarely has so great an accomplishment been so widely ignored'.[9] Weinberg's paper was not even cited elsewhere in the literature until 1970. There were a variety of reasons for this reserve. Theories with massive gauge bosons were notoriously tricky to handle and they had frequently been found to be plagued with unrenormalisable infinities. Although both authors expressed pious hopes that the use of spontaneous symmetry breaking to produce masses would remove this difficulty, it all looked very suspect in the absence of a proof. An essential phenomenological prediction of the theory was that there should be neutral as well as charged weak currents (p. 104). Neutrino experiments in the sixties had failed to reveal that effect. The experiments were notoriously tricky because the neutrinos interact so very infrequently. The average 1 Gev neutrino can traverse several million miles of lead before it manifests an interaction. In such rarefied circumstances the experiment is very vulnerable to contamination by spurious background effects. In the CERN neutrino programme it was necessary to estimate whether stray neutrons might not occasionally mock up the appearance of a neutrino interaction. It was the opinion of the experts in the sixties that all their candidate neutral current events could be explained away in this fashion. The theoretical uncertainty about consistency combined with experimental apparent noncompliance with prediction, conspired to make the Salam—Weinberg theory seem to be no more than overingenious speculation when it first appeared on the high-energy physics scene. Perhaps even Salam and Weinberg had their misgivings, for they confined their original papers simply to the weak interactions of leptons, an obviously unsatisfactory restriction.

15 *Kiev 15*

The 26th of August to the 4th of September 1970, saw the High Energy Physics Conference back in Kiev, a city of greatly enhanced prosperity since the previous visit in 1958. The meeting was now wholly conducted in English, but of the eight speakers specially invited to address the Conference in addition to the rapporteurs, seven were Russian. Such an imbalance was not altogether unreasonable since many Soviet physicists had little chance to travel outside their own country and to present their ideas in person to their colleagues in the West. Kiev 15 was a showcase for them of which they were eager to take advantage.

The general impression of the Conference, both from memory and from rereading the Proceedings, was of a vast mass of undigested material. Almost every rapporteur began his talk with an apology for having had to neglect many of the plethora of papers submitted to him. The upshot was a meeting overloaded with detail, heavy and unexciting for its participants. All in all, it was a somewhat turgid occasion.

Astier was the rapporteur dealing with boson resonances. He started with a Baconian characterisation of the experimentalist's task: 'The mission of the experimentalists is to extract from the world of their experiments all possible information and to present them to other people, in particular to theorists, with whom they have to work.'[1] The scene was confused. The A_2 argument rumbled on, with one group claiming an eight-standard-deviation effect showing splitting and another a five-standard-deviation effect showing a single peak. Such situations tend to produce declarations of loyalty rather than cool analysis. Astier fulminated against the impropriety of this behaviour. 'We are not dealing with beliefs, we are not theologians [an implied apposition of scientific reason to theological irrationality which I would not accept!], but sain [sic] men. I think the way to eliminate this bad situation is to be always aware that our results depend upon the technique and formalisms we use ....'[2] In other words, we necessarily view the physical world from a perspective; let us be conscious of the limitations our vantage point may impose.

130

(Notice that Astier is defending physical realism whilst acknowledging that the physical world is viewed in context.) His remarks and assessments did not give pleasure to all. In the discussion, Maglič (the leading proponent of the split A_2) said, 'I propose that the Rapporteurs' talks be abolished.' He wanted a more highlighted approach. 'Without this change our High Energy Conference will become boring events . . . Our job is not to test theories . . . Our role is to explore.'[3] When one seeks to assess the meaning of such an interlude one must bear in mind that the dust does settle, the material is finally digested, agreement is eventually attained. In the end the A_2 was found to be a simple resonance like everything else.

A rapporteur with a happier task was Richard Wilson, who had to cover the field of lepton–hadron interactions and quantum electrodynamics. He claimed to be 'the luckiest rapporteur at the meeting for I believe I have the most exciting data to present.'[4] For the first time for several years quantum electrodynamics was in unambiguously good shape. Various little niggling discrepancies had proved to be transient blemishes rather than permanent problems. Bjorken scaling was looking very good (the Russians had their own name for it: 'automodelity') and results of the first Drell–Yan type reaction were presented at Kiev 15. These two theorists from SLAC had suggested a mechanism for producing lepton–antilepton pairs in hadronic collisions. It involved the annihilation of a parton from one hadron by an antiparton from another hadron to produce a high-energy virtual photon capable of materialising as the lepton pair. The measured cross-section looked right but it had a rather odd 'shoulder' modifying its decrease at high energies. The significance of this shoulder would require much better resolution in the measurement before it could be elucidated.

Results from the first storage rings were beginning to come in. The collision of a beam of electrons with a counter-rotating beam of positrons produced what Wilson regarded as the 'most exciting part of this talk'.[5] At the time, energies were only high enough to exhibit the production of resonances like the ρ and ω. It was the start of an extremely powerful method for interrogating nature. Results from e^+e^- annihilation have played a very significant role in the development of high-energy physics. They will continue to do so when LEP (p. 6) comes on the air in the near future.

Quantum field theory figured more largely at Kiev 15 than in immediately previous Conferences, partly because it was a major concern of Russian theorists. The influence of Bogolyubov had something to do with this. The rapporteur, however, was Swiss. Klaus Hepp disarmingly began his talk by saying 'According to the rule that everyone gets promoted to his level of incompetence I have the honour to report on recent results in quantum field theory in perturbation theory, in axiomatic and in constructive quantum field theory.'[6] The real interest, however, lay in developments of relativistic quantum mechanics which were abstracted from quantum field theory or went beyond it in some way.

In the field of current algebra there was an increasing understanding of anomalies. It had been realised that the most naive expectation of how currents would behave was not always correct. The reason lay in the singular nature of quantum field theory. One aspect of this behaviour was the infinity problem which renormalisation theory dealt with. Another aspect manifested itself in current algebra where it produced 'anomalous' extra terms in the commutators. Sometimes these anomalies were disastrous, for they destroyed a desirable structure. In that case it became a constraint on theory choice to find situations which were anomaly-free. (One of the reasons for the extreme difficulty in constructing a consistent quantum theory of gravity is that quantised general relativity is highly susceptible to undesirable modification by anomaly.) On the other hand, anomalies can sometimes be the salvation of a theory. A powerful technique had been used for several years called PCAC (partially conserved axial current). It was a limiting argument which exploited the smallness of the pion mass in relation to other hadronic masses. Applied naively to the neutral pion it implied that the latter would not be able to decay. This disaster was avoided by recognising that the naive result was modified by an anomaly, whose presence had been identified by Steve Adler. The anomaly came from a process involving a triangular quark loop (that is a Feynman diagram with three virtual quark lines), at one of whose vertices the pion was attached and at the other two vertices was the pair of photons into which the π^o was to decay. The magnitude of this contribution (and hence the lifetime of the π^o) depended on how many different types of quark could circulate in the loop. If you just took plain quarks the answer came out nine times too small, but if you supposed the quarks to be tri-coloured the answer came out just right! Here was a valuable and direct argument in favour of colour.

Current algebra also made its own contribution to deep inelastic physics. It was recognised that the scaling regime could be looked at from a commutator point of view, but with the commutators evaluated, not at equal times as previously, but at points which were on each other's lightcones (that is, they were points between which massless particles could propagate). Lightcone algebra (together with an associated technical idea called operator product expansions) provided a sophisticated way of thinking about matters addressed in more down-to-earth fashion by the parton model. The latter had the advantage, however, that it could also discuss phenomena, such as e^+e^- annihilation and the Drell–Yan process, which eluded the grasp of the apparently more highbrow technique. These varieties of exploration of deep inelastic physics were motivated, partly no doubt by personal desires to be different, but also by the search for as wide and flexible an approach as possible to the understanding of the phenomena.

Duality was more an extension of relativistic quantum mechanics than a derivative branch of quantum field theory. Veneziano commented that it 'is indeed a very young theory, still looking for a shape of its own and a direction in which to develop'.[7] It was recognised as involving a huge degeneracy in the number of physical states. (If you are going to build up high-energy behaviour out of resonances you need an unconscionable lot of them to do so.) The string picture was beginning to attract attention.

The rapporteur for weak interactions was V.M. Lobashev. He referred in passing to those 'neutral currents whose absence is postulated by the available "conventional" models of weak interactions'.[8] Salam and Weinberg cut no ice at Kiev 15. Part of the reason was the very precise limits which could be set on the presence (or rather, absence) of neutral currents in strangeness-changing weak interactions. These decays provided what appeared to be a very sensitive probe. For example, Lobashev reported that the branching ratio (that is, the proportion of instances) for K_2^0 decay into a muon pair was less than 6×10^{-9}.

It so happened that the solution to this problem had been propounded earlier that very year. Glashow, Iliopoulos and Maiani (GIM for short) had shown that such unobserved decays could be cancelled out. There was, however, a price to pay. It was the existence of a fourth quark carrying its own particular quantum number, given the ridiculous name of 'charm'. When charm had first been proposed in the mid-sixties (p. 110), its principal purpose had been to complete a quark−lepton analogy. The charmed quark now postulated was a kind of cousin to the u, for both carried charge 2/3. GIM achieved a satisfactory way of extending Salam−Weinberg ideas to encompass hadronic weak decays. Their work was really the final stage in the articulation of unified electroweak theory. In due course Glashow was to share a Nobel Prize with Salam and Weinberg, for earlier work he had done on unification.

I remember reading GIM's paper at the time and feeling it preposterous to invoke a new quark and a new quantum number to explain away a difficulty in a highly speculative theory. Ah well! Real insight is always in short supply. Fortunately, these matters are not settled by popular vote but by the way things actually are.

16 *Chicago 16*

The 16th Conference on High Energy Physics met at the National Accelerator Laboratory, south of Chicago, from September the 6th to the 13th 1972. The synchrotron located there had been brought into action that year and was to prove capable of attaining 400 Gev. Its design and construction had been the brainchild of the laboratory's director, Bob Wilson. He had the reputation of being a go-getting corner-cutter — he is reported to have said that if a machine worked when it was first switched on, that just showed that it had been overdesigned. There had been some precarious moments in the execution of the NAL project. A similar type of machine was being built at CERN where a different, belt-and-braces, philosophy prevailed. Fermilab (as NAL came to be called) beat the Europeans to it, but not without some heart-stopping moments on the way. The Americans had particular trouble with the reliability of their magnets. CERN was not completely upstaged, however, for the ISR (Internal Storage Rings), a proton—proton collider, had come into operation there in 1971, providing access to hadronic interactions at higher energies than were available anywhere else, outside the dubitable realm of cosmic-ray physics.

Bob Wilson was something of a polymath, including architecture among his many interests. The main building at Fermilab, with its central tree-planted atrium, made a striking feature rising above the Illinois plain. Another Wilsonian imaginative gesture was to use the considerable area within the circumference of the synchrotron ring to recreate in miniature the environment of the original prairie, complete with a small herd of buffalo.

About a thousand participants gathered at Chicago 16, although some Russian invitees, like Gribov, were yet again unable to be present. The Proceedings were published in four small but thick volumes, totalling more than 1800 pages of fine print, garnished with futuristically designed covers and embellished with photographs. The Conference itself was preceded by a ten-day 'workshop', a device which enabled the discussions at the plenary sessions to be ruthlessly abbreviated. Those who were not officially invited could watch the proceedings through a television link to an overflow hall. In addition, the plenary sessions and some others were videotaped and so could be made available subsequently to an even wider audience. The 'Rochester' Conferences had become really big business.

Phase shifting had become a fantastic taxonomic industry, making use of subtle constraints derived by Cutkosky from analyticity requirements, in the effort to eliminate some of the ambiguities inherent in its procedures. One of the rapporteurs concerned was Claud Lovelace, then at the height of his powers and eccentricity. He began his report by saying, 'Most talks start with apologies to contributors **NOT** mentioned. However, if some participants knew what I **AM** going to say about them, they would have stolen my manuscript and burnt it.'[1] He clearly viewed the phase-shift activity as a craft with its own mystery: 'sophisticated mathematical methods can never be a substitute for commonsense and care.'[2] Lovelace produced a 'mathematical model' for the whole enterprise − 'Accelerator program committees are represented by random number generators. They order haphazard measurements in a feasible regime'[3] − and he characterised the use of forward-dispersion relations as producing 'an inexhaustible paper mine'.[4] There was much comment full of insight amid the pyrotechnics. 'Extrapolating the function onto its boundary is harder. Analytic mapmakers put sea serpents on the coast, not dragons in the interior.'[5] − a beautifully phrased way of making a serious mathematical point. After confessing that he did not believe in real quarks, Lovelace ended with some general observations. 'I feel I ought to mention the sociological problems of resonance physics. The experimentalists by now must feel like ants, or like pharoah's slaves building the pyramid .... What hurts most, I think, is that the pyramid is not anonymous. The resonances come out of the computer with someone else's name on them.'[6]

It was an astonishing bravura performance, the climacteric of Lovelace's public career. After Chicago 16 he gradually subsided from view as far as high-energy physics was concerned.

Another resonance session recorded the demise of the split A_2. Diebold was charitable in his comment: 'There is a lesson from statistics to be learnt from this puzzle ... low statistics experiments must be treated very carefully.'[7] In other words, don't push your luck.

The burgeoning complications of the resonance scene might look more like chemistry than physics, but elsewhere promising new ways were being developed to analyse aspects of high-energy hadronic collisions. Feynman (using his parton model) and Yang and others had drawn attention to a phenomenon called limiting fragmentation. In a very high-energy collision there tend to be many particles created, as part of the energy available is

Geoff Chew lays it on the line at Chicago 16.

transformed into matter. Rather than trying to look at all that detail, one could take a leaf out of the electron-scattering book and study inclusive cross-sections (p. 125). In other words, one could concentrate on seeing some particular sort of entity in the final state and then sum over (experimentally that meant, neglect to investigate) all else that was going on. Limiting fragmentation concentrated on a particle which carried off a significant fraction of the momentum available in the centre of mass − in other words, which could be thought of as being a substantial fragment of one of the incident particles. The suggestion was that the behaviour of this inclusive cross-section would, at high-energy (s), become independent of s and depend only on x, the fraction of momentum carried away. This was indeed found to be the case. The proper theoretical way to think about this was discovered by Alfred Mueller in 1970. His work showed the continuing fertility of the general ideas of analyticity and of Regge theory. The inclusive cross-sections were identified as appropriate discontinuities of multiparticle amplitudes, whose high-energy behaviour was, in turn, specified by Reggeism. The limiting-fragmentation result was due to the dominant effect of the Pomeron, which provided a factor of s just sufficient to cancel a kinematic factor of s^{-1} and give an energy-independent behaviour. The scheme was a neat new synthesis of quite old ideas.

The notion of looking at inclusive cross-sections was an invigorating and refreshing development for hadronic physics. The prior feeling had probably been that they were too simple to be interesting. From an experimental point of view it simply looked like seeing what beams of energetic secondary particles would be available after an initial collision. In actual fact it was only the simple situations which were really interesting, for it was only in their cases that theory was capable of making statements both clear and yet non-trivial. The organisers of the high-energy collisions session at Chicago 16 (Fox and Pilcher) were right to say 'the inclusive era is upon us'.[8] That was a great bonus for a collider like the ISR, since inclusive cross-sections were precisely the quantities most readily measured at such a facility. Later, the notion would be combined with deep scattering in purely hadronic physics to lead to further fruitful investigations. At Chicago 16 it was just beginning to be realised that parton models might have something to say about such wide-angle scattering. Henry Abarbanel was judiciously cautious in his report. 'The whole of this intriguing idea is in its formative stages and any judgement as to its lasting merits would be more than a trifle hasty now.'[9]

Duality theorists continued labouring away. They had come to realise that to avoid ghostly troubles and inconsistencies it was 'best' to formulate the theories in higher space-time dimensions (10 or 26 were the current preferences). The chairman of the relevant session, Harry Lipkin, proposed as its motto: 'Dual theory should be presented in such a way that it becomes understandable to non-dualists. At least as understandable as East Coast theories are for West Coast physicists, and vice versa.'[10] He was referring to a recognisable difference, to be found as much in physics as in general life and culture, between Californian free-wheeling (bootstrappers) and New England sobriety (field theory). The session showed that it was beginning to be recognised that the best way to attain the clarity desired by Lipkin was by appeal to the 'rubber string picture'.

Meanwhile the mathematically rigorous investigators of quantum field theory continued on their austere and lonely way. They had become decoupled from other theoretical physicists, to whom they often gave the impression of being a mutual admiration society. The report of their session is full of phrases like 'spectacular success', 'remarkable', 'decisive' and so on, but their impact was small outside the circle of enthusiasts, and it has remained so. Murph Goldberger once said of this activity, parodying its mathematical style of expression, that given $\epsilon > 0$, however small, the strict field theorists' contribution to physics had been less than ϵ.

Yet important developments were happening in basic field theory, pioneered by those who were a little more venturesome. The crucial idea was that of the renormalisation group. Its aim was to investigate how interactions behaved when they were sampled on different length scales (equivalently, momentum transfers). As a very crude example of the sort of consideration involved, think of an atom. Viewed from a distance it is electrically neutral, but if one penetrates the screening cloud of electrons one encounters the positive charge of the nucleus. Its perceived properties depend on the distance involved. The vacuum in quantum field theory acts as a polarisable medium (p. 11) and so screening effects are always relevant to elementary particles. The strength of their effective interaction will vary with the distance involved in sampling it. The renormalisation group is a mathematical way of coming to grips with this. Once again, high-energy physics had something to learn from another subject. Statistical mechanics had had to grapple with such problems, particularly in relation to the discussion of phase transitions (such as melting) in which long-range order can be established or broken. Ken Wilson had worked in both elementary particle physics and condensed-matter physics and he brought a powerful mind to bear on these problems. Some surprising techniques were involved, such as analytic continuation of the dimension of space-time. Wilson, later a Nobel Prize winner, gave a difficult talk at Chicago 16 about these techniques 'stolen from previous work in statistical mechanics'.[11]

There was a session on quarks and all that, organised by Gell-Mann and his young collaborator, Harald Fritzsch. There was about it an inescapable air of uneasiness. Quarks seemed to work so well, but where were they?

Lightcone algebra was a device to try to dodge the issue, to get the benefits without the embarrassments. 'For more than a decade, we particle theorists have been squeezing predictions out of a mathematical field theory model of the hadrons we don't fully believe.'[12]. Then there was the perplexity of all the different points of view on offer — partons, bootstrap, duality, even a colour gauge group theory (whose possibility was noted in passing as being an 'interesting question'[13]). In the end it seemed best to have a bet on every horse in the race. 'Since the entities [quarks] we start from are fictitious, there is no need for any conflict with the bootstrap or conventional dual model view.'[14] Meanwhile, the quark search continued. Adair compared it to a girl about to get married who ordered seventeen yards of material for her nightdress. She was going to wed a particle physicist, for whom the search was as important as the discovery.[15]

But above all, Chicago 16 was the Conference at which the electroweak unification theory received recognition as an important advance in the understanding of fundamental physics. Ben Lee said, 'Perhaps the most significant development in weak interaction theory in the last two years, both from the viewpoint of theory and of possible impact on future experiment, has been the construction of a renormalisable model of weak interactions based on the notion of a spontaneously broken gauge symmetry.'[16] It was at a European conference in 1971 that this previously conjectured possibility was perceived to be an actual reality. 'At the Amsterdam conference last year, a young Dutch physicist, G. t'Hooft, not yet out of graduate school, presented a paper which would change our way of thinking in gauge theory in a most profound way.'[17] It was an astonishing achievement by this young man to have cracked the problem of renormalisability. The only comparable debut I can think of in recent physics was Brian Josephson's discussion of superconducting junctions, which he too produced when a graduate student.[18] The Salam—Weinberg theory was now theoretically respectable and, just as happened with quantum electrodynamics when its renormalisability was discovered, this opened up the chance to perform many more credible and detailed calculations. Weinberg recalled that, 'As soon as it was realised that spontaneously broken gauge theories are renormalizable, there was a great explosion of theoretical effort devoted to detailed calculation of higher-order weak and electromagnetic "radiative" corrections and to the construction of alternative models.'[19] In the course of this further investigation a very interesting feature emerged. In order not to spoil the renormalisability of the theory it was essential to remove a certain anomaly (p. 132) which otherwise would have frustrated the consistency of the renormalisation procedure. This could be done by cancelling lepton contributions to the anomaly against those of the quarks. Only if there were the three-fold quark degeneracy required by colour would this work out.[20] In a most delicate and unexpected way the electroweak theory served to confirm the hypothesis of coloured quarks. One had the feeling that theory was really on to something. The main difficulty for electroweak remained the question of neutral currents. The experimentalists were already beginning to reconsider the matter. Meanwhile, not all theorists had climbed aboard the bandwagon. Bjorken, in a talk at Chicago 16, said of spontaneously broken theories, 'I don't know any 'believable' theories . . . . For me, a believable theory of this class will necessarily lead to at least some gain in understanding Great Questions.'[21] He set his standards very high. By 'Great Questions' he meant something like why the muon and the electron have the masses they do.

Such caveats notwithstanding, when Murray Gell-Mann came to give the Conference summary talk he felt able to award the electroweak theory his palm of honour. 'I should like to express my enthusiasm for recently revived attempts to unify the electromagnetic and weak interactions in the framework of a renormalisable theory.'[22] It corresponded to the intuition — widely held in the physics community by many others in addition to Gell-Mann — that 'Above all, the theory should be beautiful.'[23] Murray drew a zig-zag hump, which he said was not an experimental histogram but a Babylonian ziggurat. The variety of theoretical ideas on offer was being compared to Babel, yet 'They are all, or nearly all compatible if we can get our *language* straight.'[24] Gell-Mann had a deep interest in linguistics and he was taking a Chomskian view of an inherent structure underlying the varieties of theoretical language. He did not, however, omit his customary caution that 'Real quarks, detectable in the laboratory, are not required by theory.'[25]

140

17 *London 17*

I was on the Organising Committee for the 17th Conference, held in London at Imperial College, from the 1st to the 10th of July 1974. Our job was to try to arrange the programme of sessions to cover the ground without too many clashes of interest during the parallel sessions. The detailed logistical arrangements were in the competent hands of the staff of the Rutherford Laboratory. An attempt was made to encourage the rapporteurs at the plenary sessions to respect the limited time assigned to them and so end the abuse of overrunning which had become tiresomely frequent at recent Conferences. Each speaker was offered a bottle of good claret if he was punctual in his presentation. All but one succeeded in keeping to time. In the event, the committee was not hard-hearted enough to withhold the bottle allocated to the sole, and not very extensive, offender.

The Conference reception was held at the Royal Academy, whose walls carried the paintings of the Summer Exhibition. I regret to record that there were some participants who complained that they did not get enough raspberries and cream. The adequate and swift service of refreshments was not an outstanding English capability. A few years earlier there had been an international physics conference at Oxford, whose organisers had the thought that an ox-roast by the Thames would be a suitable celebration. The beast provided was big enough but it had not been realised how long it would take to slice it up. I recall a Breughelesque scene in which hungry elementary particle physicists pressed round the harrassed cooks, the physicists' ravenous faces lit by the flames of the fire as they jostled each other to get to the meagre supplies available. The Oxford ox-roast took its place with the Niagara Falls trip and the Dubna boat ride in the annals of shared adversity, preserved in the folk memory of the high-energy physics community.

The Proceedings of London 17 had a novel feature. Beautifully printed and promptly produced, there was only one snag. The glue bonding the pages together soon deteriorated to produce a crumbling volume. This autodestruct quality was perceived by some to have a certain appropriateness for a subject in such a constant state of flux.

The two years surveyed by the Conference had been an exceptionally interesting and fast-moving period. In 1973 the CERN experimentalists beat the Americans to the discovery, at long last, of neutral currents in weak interactions. From then on the quantity of accumulating data proved consistent with the 'standard model' of Salam and Weinberg. The dust had really settled for electroweak unification.

A great advance had also been made in understanding quark theory. In 1973 Gross and Wilczek, and independently Politzer, discovered that a colour gauge theory could be asymptotically free. This was a splendid result, arising from renormalisation group investigations (p. 138). Recall that the effective coupling constant depends on the region over which it is sampled. Asymptotic freedom is the property that when that region becomes very small the coupling tends to zero. In other words, in the point-like limit, quarks in a colour gauge theory behave as if they were freely rattling around and not tightly bound to each other. That is exactly the situation needed to make sense of the parton picture of deep inelastic scattering (p. 126). Only gauge theories have this property.

If the coupling decreases at small distances, it will increase at large distances. It was natural to go on to hope that this increase would be unbounded, that the other side of asymptotic freedom would be 'infrared slavery'. Infrared corresponds to large distances, and the slavery would result from a limitless growth of the force between quarks with increasing distance. It would be as if there were a perfectly elastic string coupling them together. That way one would have confinement − the quarks would never be able to escape however hard they were struck in a deep inelastic collision. One could have one's cake and eat it: all the benefits of the quark parton model, plus an explanation of why no free quark had ever been seen. If that were so, the quark level in the structure of matter would exhibit a property never previously encountered. All earlier constituent pictures had found their final experimental verification by producing the separate bits they claimed as constituents − electrons stripped from atoms by ionisation; protons and neutrons ejected from nuclei in nuclear reactions. If confinement is true, this will never be possible for quarks.

The property of confinement remains a conjecture. It has not been proved to be the case. The relevant calculations are much more difficult than those which established asymptotic freedom. Among the lines of attack which date from this time, and which are still continuing, is lattice gauge theory. This replaces continuous space-time by a discrete set of points and it searches for insight by trying to find out what happens in this greatly modified case.

It often seeks to draw on analogies from statistical mechanics, a subject much experienced in the discussion of crystal lattices. It also demands powerful computing facilities for its implementation.

Meanwhile, more phenomenological exploitation of the quark theory continued apace. A model was developed at MIT which was given the technical name 'bag'. It was both crude and clever, a relativistic way of formulating the naive picture of quarks free to rattle around inside the impenetrable walls of a container. In explaining the spectrum of hadron states it scored some success.

The ideas of deep inelastic scattering were extended to purely hadronic processes. From 1973 onwards important results became available from the CERN ISR of the inclusive cross-sections for the production of large transverse momentum (p_T) hadrons at high-energy (s). The notion that these rapidly sideways-moving particles were fragments of entities involved in a primordial hard scattering led to a scaling law. The cross-section was predicted to be proportional to an inverse power of s multiplied by a function of the dimensionless ratio p_T^2/s. (This was the analogue for this case of the Bjorken scaling in leptonic processes.) The exact power of s depended on the nature of the entities involved in the hard scatter (that is, the initiating internal interaction transferring a large amount of momentum). Quarks gave an s^{-2} decrease, but since the observed behaviour then accessible appeared to decrease more rapidly than this, it was necessary to consider various other hard scatters in which hadronic fragments were substituted for quark constituents in the initiating interaction. A great industry arose, particularly at SLAC, Caltech, CERN and Cambridge. The eventual outcome was, as always with hadronic physics, pretty messy. Yet certain salient features were clear. Scaling behaviour was indeed present. As one was able to move into higher-energy regimes, the s^{-2} contribution of quark scattering emerged and became increasingly important. Confined quarks could not themselves appear solo, but they generated observable jets as their energy materialised in the acceptable form of hadrons travelling in the direction of the initiating sideways quark, of which they were fragments.

I worked in this area for several years with my friend and colleague, Peter Landshoff. I would like to recall one minor but satisfying triumph. It did not concern inclusive processes but rather the wide-angle elastic scattering of protons, another process with a large transfer of momentum but, in this case, without any production of extra particles. We figured out that it should behave at fixed scattering angle (θ) like a power of s and that this power would be the same for all values of θ. We decided to try to check this. The available data were contained in many different references, so we asked a highly computerised friend of ours whether he would let us have a printout of the combined data that he was trying to analyse using all sorts of fancy procedures. He kindly did so. We simply plotted it on log—log graph paper to give a series of parallel straight lines corresponding to the different values of θ available. This exactly corresponded to the behaviour we had expected on the basis of the parton picture. It was gratifying that a little theoretical nous yielded a simple and convincing result which had slipped through the net of more sophisticated analyses.

Fred Gilman was surely right to say that 'the ideas of scaling, the lightcone behaviour ... and quark parton model ... effect [sic] aspects of the subject matter of almost every session at this conference'.[1] In triumphant mood, he went on to say, 'Which of you thought six years ago in Vienna that ... [these ideas] ... would permit one to predict within 20% or better e, μ, ν, $\bar{\nu}$, experiments over two orders of magnitude in ν_1 and q^2?'[2] The analysis of deep inelastic scattering of leptons and neutrinos had reached that agreeable state of tractibility by London 17.

There was one area where parton ideas were relevant but which seemed to be in real trouble. It was the phenomenon of e^+e^- annihilation. At sufficiently high-energy (s), the total cross-section for annihilations in which the energy materialised as hadrons, should be given by s^{-1} multiplied by a factor, R, which is just the sum of the square of the quark charges. At moderate energies this prediction seemed to agree with experiment ($R = 2$), provided one thought of the three quarks then established in the popular mind (u, d, s) and provided one tripled the number of quark types to allow for colour. It was another success for the whole chromatic idea. Trouble arose, however, when the energy increased further.

From 1973 a joint Berkelely—SLAC team measured R in an extended region of energy up to about 3 Gev. They were using SPEAR, a large e^+e^- collider built at Stanford. It was found that R began to increase above 2. At London 17 Burt Richter reported these results. There appeared to be a grave problem for quark—parton theory. Richter, a tough-minded experimentalist, obviously took great delight in setting the experimental cat among the theoretical pigeons. He noted that his session had only five experimental papers but 61 theoretical contributions offering more or less fanciful explanations of this unwelcome behaviour. The actual explanation was much more interesting than any of those proposed.

It began to emerge after London 17 in an episode which acquired the title 'The November Revolution'. It involved in a most curious way American East and West Coast groups working on two totally contrasting experiments. At Brookhaven a group led by Sam Ting was taking a second and more careful look at the Drell—Yan production of lepton pairs (p. 131). Theirs was a synoptic experiment, surveying a wide range of effective energies

144

and so capable of seeing anything that was there, provided their experimental resolution was fine enough not to smear it out beyond recognition. One might say that they had put on their glasses to see if they could spot any needles in the haystack. The other experiment, at SLAC, had no problem about resolution, but so fine was its net that they had an awful lot of meshes to worry about. One might say that they were examining the rick straw by straw, to see if one blade turned out to be a needle. The East Coast group had to be sure they'd really seen something; the West Coast group had to be lucky enough to look in the right place. By an extraordinary coincidence, on 11th November 1974 both groups decided that they had come across a striking new phenomenon. Ting's crowd had resolved the Drell−Yan shoulder into a very narrow resonance spike at 3.1 Gev energy. Richter's crowd found that as they tuned the SPEAR energy through 3.1 Gev all hell broke loose in their detectors, due to a plethora of events originating from the same resonance. The East Coast group had discovered what they called J. The West Coast group had discovered what they called ψ. For a while there was a rather silly nomenclature war between the two groups, even with t-shirts being handed out emblazoned with J and the details of the resonance. In the end the tactful compromise of J/ψ entered the literature. A few days later SLAC found a second resonance, ψ' at 3.7 Gev.

The discovery removed the R problem. The 'linear rise' had been an artefact of the presence of the resonances. It left, however, a striking new problem in its place. Pais comments, 'The general pandemonium following these discoveries compares only, in my experience, to what happened during the parity days of late 1956.'[3] The problem was the widths of the resonances. They were very narrow. That meant that each resonance was very long-lived. Yet each was also very massive and so would be expected to decay rapidly. What stopped it doing so? All sorts of ideas were proposed, but in the end a most remarkable explanation gained universal acceptance. It depended on taking the existence of charm (p. 133) absolutely seriously. I have described the explanation elsewhere.

> The way it worked was peculiarly subtle. The J/ψ is of zero charm but it is made up of a charmed quark c and a charmed antiquark $\bar{c}$ which stay far enough apart within J/ψ not to annihilate each other straight away. Of course their equal and opposite charms cancel out, leaving J/ψ charmless. In a word, it has hidden charm. But why then should this stop the admittedly charmless J/ψ decaying very fast into ordinary particles like pions? The frank answer is that we don't know, but we have seen this sort of thing happening before. There is a meson resonance called ϕ (phi) which is a state of hidden strangeness made out of s and $\bar{s}$. However, it does not decay into pions of zero strangeness but much prefers to decay into a K meson and a $\bar{K}$ meson,

145

that is into a strange particle−antiparticle pair. . . . Presumably, the same is true, *mutatis mutandis*, for J/ψ. It would very much like to decay into a charmed particle−antiparticle pair. However, particles with non-zero charm turn out to be rather heavy and the J/ψ does not have enough energy to decay into a charmed particle−antiparticle pair. It is like a man with £1.90 in his pocket who believes it is vulgar to tip with less than a note and who consequently cannot reward both the cook and the butler. This impasse holds the J/ψ impotent until eventually it swallows its pride and turns into non-charmed particles.[4]

Of course, when one went to somewhat higher energy one would expect to observe the production of particles with explicit charm. In due course that was verified and this provided the final vindication of the charm idea.

All this was in the future at London 17, but there were some there who were confident that charm, in one form or another, would figure on the eventual agenda. In an exuberant talk John Iliopoulos asserted, 'I have already won several bottles of wine by betting on neutral currents and I am now ready to bet a whole case that if the weak interaction sessions of this conference were dominated by the discovery of neutral currents, the entire next conference will be dominated by the discovery of charmed particles.'[5] He had surveyed the whole history of weak interaction theory, commenting about the $V - A$ theories of the 1960s that, 'Once more, when physicists were confronted by a consistent theory [a contrived theory, then known to be renormalisable in IVB terms] and an elegant theory [$V - A$, not then known to be renormalisable in those terms] they unanimously chose the latter'[6] − and were right! Iliopoulos gave an extremely lively talk, to which the cold print of the proceedings cannot do justice.

More old-fashioned physics (at low momentum transfers) made experimental progress without occasioning any remarkable theoretical breakthrough. Total cross-sections were found to rise with energy, a result which is in qualitative accord with the expectations of the Reggeon calculus, although the latter remained extremely difficult to handle in a convincingly detailed way.

Resonance physics − now 'low-energy physics' in these conferences obsessed with extreme high-energy regimes − ground on. Peter Litchfield concluded his talk on baryon resonances by saying, 'I would compare this review to a striptease. I hope it will be suggestive, but you may be left with the impression that there is something you have not seen.'[7] In more serious mood, Rosner said, 'We have a Balmer spectrum; some even claim a Bohr theory. Now we are waiting for quantum mechanics to be invented.'[8] He was referring, of course, to dual theory. The string picture was developing and it had been recognised that, in appropriate limits, it yielded gauge theories and general relativity. This encouraged David Olive to say, 'Originally designed for hadrons, the dual theory seems capable of describing weak, electromagnetic and gravitational interactions. Thus we have the tantalising possibility that there may emerge a unified theory of all interactions.'[9] If superstrings do indeed prove the new 'invention of quantum mechanics' by being a Theory of Everything, Olive's words will have proved truly prophetic. It was only in the 1980s that this programme really got under way.

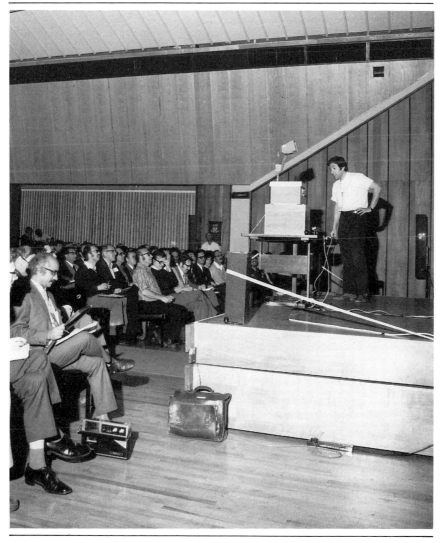

John Iliopoulos makes his bet at London 17.

147

From this time onward the search for a GUT (Grand Unified Theory) is definitely on the speculative agenda. The initial attempts made no use of string theory but sought to imbed $SU(2) \times U(1) \times SU(3)_c$ (that is, the combination of standard electroweak with quantum chromodynamics, the gauge theory of coloured quarks) in some larger, all-embracing, group. Such an endeavour inescapably mixes leptons and quarks and so imperils the stability of the proton, implying the eventual decay of all normal matter. Don't worry, though. The decay of a proton involves disposing of three quarks. In consequence, the proton is very long-lived, lasting on average more than 10^{20} times the present age of the universe. That will see us through. Skilful searches for proton decays occurring as extremely rare events in large quantities of matter have so far proved fruitless. The null results of these experiments place quite strong constraints on the possible forms of GUTs. (Serendipity in physics is beautifully illustrated by the same detectors having picked up neutrinos from a recently observed supernova explosion!)

A further potentially fruitful idea discussed at London 17 was supersymmetry, invented by Johannes Wess and Bruno Zumino. They had the ingenious notion of invoking symmetries which interchange bosons and fermions. Previously, people had only considered symmetries which did not mix up these two families of particles. Crudely speaking, the more symmetrical a theory is, the more constrained it is, and the more controllable and limited its infinities and anomalies are in consequence. That raises the hope that the supersymmetry idea might even bring tractability into quantum theories of gravity (supergravity), although the current best hope in that respect is held to be superstrings.

The final Conference summary at London 17 was given by the Italian theorist, Sergio Fubini. He said that in his opinion 'the most fundamental fact which has emerged during the last decade is that a fully symmetric Lagrangian can give rise to non-symmetric results'.[10] He was referring to spontaneous symmetry breaking, which he illustrated by reference to the fourteenth-century story of Buridan's ass. The unfortunate animal is placed equidistantly between two bundles of hay, one to his right, the other to his left. If he is not to starve to death, the animal must break the symmetry of his situation by moving to one side or the other. Fubini spoke about supersymmetry, saying it was necessary that 'the whole scheme should not make Sidney Coleman unhappy'.[11] Coleman, a mordant East Coast theorist, had played the role of mathematical conscience to high-energy physics by his critical strictures on sloppy theories, earning for himself the sobriquet of 'No-go' Coleman.

Fubini had been a important figure in my generation of theoretical physicists. His London talk had, perhaps, more style than content. The world of high-energy physics was changing very rapidly and those in their forties, whose recollections and experience went back to the beginnings of the Rochester era, were beginning to find themselves a little out of breath with it all. It affected the best of us. From now on, even Murray Gell-Mann, though certainly not inactive, will not have the dominating role that has made him the hero of my tale so far.

18 *Tibilisi 18*

The 1976 Conference on High Energy Physics was held in Tibilisi, in the Georgian Republic of the Soviet Union, from the 15th to the 21st July 1976. About 1200 papers were submitted to it and around 750 delegates attended its 25 parallel and 15 plenary sessions. Professor A.M. Petrosyants, Chairman of the USSR State Committee for the Utilisation of Atomic Energy (a subject very remote from elementary particle physics as practised in the 1970s), sounded a cautionary note in his opening address. The subject and its community were both growing apace. 'In this connection we consider it important that you could, when summing up the results and outlining new problems, reasonably and critically commensurate your results with the expenditures a society can permit for the physical investigation of the fundamental properties of matter.'[1] The taxpayer's pocket must not be assumed to be inexhaustible, a message being uttered worldwide.

In many ways Tibilisi 18 was a consolidatory conference at which the main features of the quark theory of matter were acknowledged as having been satisfactorily established. Bag modellers were working away at their attempts to calculate the consequences for hadronic states, with the Soviet physicists speaking up for their version 'which we now call the "Dubna bag"'.[2] Large p_T interactions (see p. 143) were being analysed successfully in quark−parton terms and evidence began to accumulate that they had the expected two-jet character − that is, the transverse momenta balanced out by there being two groups of oppositely sideways-moving particles, which could be interpreted as fragments of the two entities emerging from the internal hard scattering, which was held to be responsible for the production of the large p_T. These results on large transverse momentum were described as 'one of the major achievements of high-energy physics over the past two years'.[3]

The situation looked even better from a more fundamentally theoretical point of view. The Spanish theorist Alvaro de Rujula gave a triumphant account

150

of the successful synthesis based on four quarks plus quantum chromodynamics (strong interactions) plus Salam–Weinberg (electroweak). A key issue here, of course, was the confirmation of the fourth quark carrying charm. This required the establishment of the existence of particles endowed with what was somewhat archly called 'naked charm', to be contrasted with the hidden charm of the J/ψ. In de Ruluja's view the message of Tibilisi 18 was that 'charm is extremely likely to have been discovered'.[4] As the year progressed it became increasingly clear that this was indeed the case. The search was conducted by an ingenious synthesis of older techniques. Charmed particles have short lifetimes (10^{-12} seconds or less). One could search for such quick decays in photographic emulsions but the problem was to know where to look. Monitoring possible decay products by the use of electronic counters enabled experimentalists to pinpoint where promising candidates events were to be found.

In his breezy style, de Rujula concluded that 'Quark believers, whether bound or unbound, can no longer be considered a bunch of religious fanatics.'[5] Quarks were emerging from the 'presumably mathematical' closet into the light of physical reality, despite the continued absence of an encounter with a single fractionally charged entity. While many problems remained in the attempt to calculate strong-interaction effects (and in particular to establish the hypothesised property of confinement), there was for the first time in postwar elementary particle physics a credible fundamental theory, quantum chromodynamics (QCD), with which to wrestle. QCD is the gauge theory of coloured quarks, whose gauge bosons are the gluons 'sticking' the quarks together to form colourless hadrons. The gluons are themselves coloured, and so are confined according to contemporary wisdom. In consequence, it is not necessary to break colour gauge symmetry in a spontaneous manner. Massless bosons which never appear on their own are quite acceptable. QCD provides an elegant fundamental theory of strong interactions.

De Rujula rightly said that 'The situation with QCD has no precedent. It is now a respectable opinion, and possibly the right one, that QCD is the correct theory of strong interactions.'[6] He went on to summarise, as 'few but important' 'Virtues of the Standard Model in the Social Sector':

    (i) The model provides an *orthodoxy*, to be (dis)proved by experimentalists and improved or superseded by theorists. Even if eventually wrong, an orthodoxy is always useful, as a reference and a framework.

    (ii) The model in the recent past has developed quasi-violent partisans and detractors. As a consequence of this we have had and have an awful lot of fun.

    (iii) The model justifies some of our salaries.[7]

A serious point lies beneath the facetiousness. 1976 is a watershed in the history of high-energy physics. Up until then, in the period we have been surveying, the subject was largely experimentally led. Theorists were seeking most of the time to respond to new discoveries as best they could. It was an era of bottom-up thinking, when particularities, rather than grand general principles, were the determinators of progress. Even so ambitious a programme as the bootstrap was in practice largely anchored in relentless attempts to explain properties of specific resonances, like the ρ. Since 1976 much of theoretical discourse in high-energy physics has been different. The possession of a credible basic theory has encouraged the search for yet grander syntheses. Top-down thinking, predicated on the basis of ever more abstract and powerful principles, has become the order of the day. The present excitement over superstrings represents a scheme whose essential phenomena lie many orders of magnitude beyond experimental accessibility and for which the Standard Model, described by de Rujula, is a kind of 'low-energy theorem', an almost trivial limiting case. While one admires the cleverness and confidence of speculation involved in such an enterprise, one also recalls that most lasting insights in physics have arisen from painful interaction between theoretical inventiveness and experimental intractibility.

One question that experimentalists had to determine for the Standard Model of 1976 was how many quark types (or 'flavours' as the unremittingly facetious called them) there actually were. In his main account, de Rujula assumed four, yet even then developments were beginning which were to lead to an extension. He had commented that 'The history of high-energy physics can naively be told, I believe, as the alternating success and failure of a principle we do not yet understand: lepton—hadron symmetry.'[8] In 1920 there had been one of each: the electron and the proton. In 1932 the discovery of the neutron broke the symmetry, but the gradual acceptance of the hypothesis of the neutrino restored it again. By 1976 one had two lepton pairs (e, v_e; μ, v_μ) and two quark pairs (u, d;c, s). At Tibilisi 18 rumours were already around that a further lepton, τ, of very high mass (1.8 Gev) had been discovered at SLAC. One might expect it to have its own neutrino (as proved to be the case). In that case, there was theoretical incentive to hope for a further quark pair to redress the balance yet again. That would have some attraction for another reason. With six quarks there was a technical possibility to incorporate *CP* violation 'naturally' into the theory. As de Rujula said, '*CP* violation is a phenomenon where counting quarks really counts.'[9]

In his conclusion, de Rujula recalled Iliopoulos's confident and correct bet at London 17 that the next Conference would be dominated by charm. He went on to say, 'I am willing to bet a *whole wine store* that the next conference will again be dominated by the discovery of *charmed particles* and their pals. The pals are an afterthought.'[10]. It was clear that he had in mind those extra quarks.

At every Conference there are items which do not attract much attention at the time but which prove to be the seeds of future fruitfulness. At Tibilisi 18 there was some discussion of 'cooling', that is, of techniques which would reduce the random scatter of the momenta in a beam of particles to make it more homogeneous. The eventual solution of the problem (along different lines from those discussed at Tibilisi 18) was an essential step in making possible the construction at CERN of the proton–antiproton collider, which in turn set the final seal on the electroweak synthesis in 1983 by confirming the existence of the intermediate vector bosons, W and Z.

One of the theoretical discussions at the Conference concerned soliton solutions in quantum field theory. These are persisting bundles of energy which the field can sustain. They represent an unexpected enhancement of the 'particle-like' states that the theory possesses, over and above those which are exhibited in the course of a standard perturbation theory analysis. Solitons require a different kind of approach. Tony Skyrme had investigated these matters many years ago at Birmingham, but his work had then been considered eccentric and it attracted little attention. Now theorists realised that it held out, in the words of the Soviet physicist L.D. Fadeev, 'a nice possibility of having a rich particle spectrum in a model with few fields.'[11] In the 1980s, the study of field-theory solutions (monopoles, instantons, etc.) has attracted a good deal of attention, not only from physicists but also from mathematicians. This has led to the solution of deep problems in topology. In fact, the 1980s have seen what seems to be the richest and most mutually satisfactory interaction between mathematics and physics since the nineteenth century. When I told my old friend, the distinguished mathematician Michael Atiyah, that I was leaving physics, he told me he had just decided to take it up.

19 *Tokyo 19*

The cycle USA−Western Europe−USSR was broken in 1978 when the 19th Conference on High Energy physics was held in Tokyo, from the 23rd to the 30th August. Japanese theoretical physicists had played an important role in the subject since the days of Yukawa in the middle thirties. Now that country had also entered the experimental field. The Conference programme included a visit to KEK, the institution operating a 12 Gev proton synchrotron. The president of the Conference was Hideki Yukawa himself, but age and ill-health prevented his actually attending its sessions.

It was a Janus-faced conference, looking backward to the completion of the standard quark model and looking forward to the opening discussion of issues which were to be the concern of the 1980s. As far as the former was concerned, the most important discovery had been that of the upsilon resonance. It was pleasing that it was made by Leon Lederman and his group at Fermilab. Lederman had been involved in the original Drell−Yan experiment, which had been unable to resolve the shoulder and so missed the J/ψ. In 1977, in the course of a further lepton-pair search, his group did come upon another very narrow resonance. The upsilon was an elder brother to the J/ψ and by now everyone knew how to interpret this state of affairs. The Υ was taken to be a $b\bar{b}$ state, where b was a new quark, characterised by a new quantum number, which for a while some called beauty but which was to end up with the more prosaic name 'bottom'. By 1980 'naked-bottom' mesons would be found exhibiting this new quantum number explicitly. Its presence went half-way to redressing the lepton−quark balance which the discovery of the τ and its neutrino had disturbed. If it were to be fully restored a further quark, t (top rather than truth), would be needed. It has not so far been found. Since no one has any idea what sort of mass top particles should have, it is not clear whether this is a matter of a pleasure postponed or whether it is a sign that once again nature knows a trick or two unsuspected by the theorists.

Of course, b quarks were fairly old hat by 1978. The rest of the

154

standard model seemed to be in good shape. Rick Field said, 'QCD is not just "another theory". If it is not the correct description of nature it will be quite some time before another candidate theory emerges.'[1] There was no general expectation that such an alternative theory would prove necessary.

The more speculative forward-looking part of the proccedings at Tokyo 19 belongs to the physics of the 1980s and it is not the prime concern of this present account. It is characterised by an ever-increasing cast of hypothetical but unobserved particles. Even the standard model makes its contribution to this list. The Higgs bosons (p. 114), invoked to bring about spontaneous symmetry breaking, have not yet put in an appearance. In a final summary talk at the Conference, Nambu expressed the hope that the Higgses are not independent particles but that they are dynamically generated − that is that they are really ciphers for some subtler symmetry-breaking mechanism, yet to be discovered.

Meanwhile other 'desirable' entities multiplied as rapidly as theoretical speculation mounted. The extension of supersymmetry ideas to gravity seemed to hold the chance of bringing renormalisability into the otherwise infinity-ridden subject of quantised general relativity. It is characteristic of supersymmetry to introduce many new partners for familiar friends. That unobserved, but universally believed-in, particle the graviton, mediating the force of gravity, would have to be massless and of spin 2. In a supersymmetric theory it acquires a spin-3/2 comrade, the gravitino. Another particle was suggested to cope with a problem arising in a different area of quantum field theory. This was the axion, whose role was to remove a potential embarrassment about the very low level of *CP* violation found in nature. Studies of the topological properties of field theory indicated that unless one was either very careful or very lucky the true vacuum state of the theory would incorporate a large propensity to violate *CP*. The axion was an attempt to provide a way around this, but it has not obliged by exhibiting itself to experimentalists.

A temperamentally bold speculator was Abdus Salam. He ended his talk at Tokyo 19 by saying, 'Some of the ideas I have spoken of above appear at this time simply crazy. . . . One should rather ask with Niels Bohr: is the idea crazy enough?'[2] That quotation from the early days of quantum mechanics (a powerfully evocative period in the folk memory of the physics community) is a dangerous one, but it certainly caught what was to be the spirit of the next ten years.

20 *Madison 20*

These accounts have tapered off as the standard quark model has become part of the accepted framework of high-energy physics. The subject itself did not taper off, but theoretical attention concentrated more and more on the speculative and still unfinished story of what might lie beyond. That is a tale for another author in another book. Coincidently, this was also the period of my own withdrawal from theoretical physics. By the time the 20th Conference convened in the USA, from the 17th to the 23rd July 1980, at Madison, Wisconsin, I was a student at Westcott House, Cambridge, preparing for ordination to the Anglican priesthood. I had greatly enjoyed my time in high-energy physics and I had no regrets for having spent a large fraction of my adult life in that pursuit. Indeed, I regarded it as having been a Christian vocation to employ my modest talents in that way. Yet I was conscious that in these mathematically based subjects most of us do not get better as we get older. There is a need for a flexibility of mind whose possession belongs primarily to one's early years. I had seen a number of my senior colleagues grow more and more miserable as they found they played a lesser and lesser part in what was going on. I was feeling increasingly intellectually breathless in the attempt to run to stand still in the looking-glass land of a subject in such rapid and continual flux. It seemed best to me to quit physics before physics quit me. Also, I was deeply interested in theological issues, which were to me even more exciting and profound than those which physics set before one. I was to find in the pastoral ministry and in the attempt to understand how the theological and scientific world views relate to each other,[1] a satisfaction greater even than I had enjoyed as one of the honest toilers in the field of theoretical physics.

In 1980 the crucial gap remaining to be filled in the edifice of the standard model was the discovery of the W and Z. This was duly plugged in 1983. Successful as this standard model certainly was, it fell short of what one might ideally hope for in a truly unified account of high-energy physics. QCD and electroweak were simply juxtaposed, without any true integration.

156

Yet there was reason to hope for a more profound relationship. As the energy is increased, the strength of the strong interactions diminishes (due to the consequence of asymptotic freedom). For electroweak, on the other hand, the strength increases with increasing energy. Eventually, they will coincide. Might one not hope that at that energy the two interactions would be perceived to be parts of a single whole, just as electromagnetic and weak interactions were found to unite at 100 Gev ? It is not hard to calculate the energy at which this grand unification might be expected. It is 10^{15} Gev, twelve orders of magnitude above energies currently attainable in the laboratory. An alternative way of expressing this elusive grand unification scale is in terms of the distance which would have to be probed to see it. That comes out at 10^{-31} m, which bears the same relation to the size of a proton as does a speck of dust to the size of the solar system. At Madison 20, H. Sugawara commented about GUTs that 'While the degree of aesthetic anxiety is reduced compared with electroweak theory by the inclusion of strong interactions, the physics content of grand unified theories is much harder to confront with observation.'[2] In fact, only in the very early universe are energies available on the required scale. Most of the discussion has concentrated, therefore, on the encounter of particle physics with cosmology in their joint attempt to describe that primeval period. Some qualitative successes have resulted, such as the suggestive explanation of why the universe is predominantly made of matter, not antimatter (p. 108). The most striking prediction of GUTs for contemporary physics is the ultimate instability of the proton (p. 148). The simplest expectation for its lifetime was about 10^{31} years, but that now appears to be in contradiction with observation. It is not beyond theoretical ingenuity to get around that, but it was certainly a disappointment to the GUT faithful to have to contrive a longer lifetime.

An extraordinary consequence of the GUT philosophy, if taken with the utmost seriousness, is that there is a desert region between electroweak unification (100 Gev) and grand unification (10^{15} Gev) in which 'nothing really interesting happens'. Of course, there will be some details to worry about, like the Higgses, but the discovery of the W and Z will have been one of the last fundamental experiments possible in a high-energy physics laboratory. In other words I have told just about the whole story of accessible experimental input into my subject. The future lies solely with the speculative insights of the theorists and the cosmologists. A sad thought, you might think. Take courage! The lessons of history are against such a view. It would be astonishing indeed if the future did not have some surprises in store.

21　*What was happening?*

The writing of this account of the development of high-energy physics over three decades has not simply been an exercise in nostalgic reminiscence, nor yet just the recounting of a tale that seemed worth the telling. I want to use it also as a means of assessing the nature of the scientific enterprise. Thomas Kuhn has emphasised the need to test ideas in the philosophy of science against examples drawn from the history of science.[1] I think that aim is admirable, even if I read that history very differently from the way that Kuhn does. In order to put the matter succinctly, Imre Lakatos has adapted an epigram of Einstein's (originally relating to science and religion, and itself parodying Kant): 'Philosophy of science without history of science is empty; history of science without philosophy of science is blind.'[2] Philosophy's void will not be filled solely by endless reconsideration of Copernicus and Galileo, or of the phlogiston controversy, but it will need to draw on many examples from the whole history of mature science.

What content, then, does the history of high-energy physics in our period have to offer to the philosophy of science? Physics is a human activity, pursued in an interactive community. It is characterised by what Michael Polanyi called 'conviviality'.[3] The Proceedings of the 'Rochester' Conferences, particularly those meetings held before the Conferences became large and formal, convey that tone very clearly. Computers may analyse bubble-chamber photographs or search for phase-shift solutions, but they will never produce those creative insights (e.g. parity violation, the idea of colour, and spontaneous symmetry breaking) which are the true rewards for all the weary labours of experimental investigation and theoretical exploration. There is no algorithm whose execution will lead to a Nobel Prize.

Because physics is a human activity there is nothing inevitably monotonic about its progress. It does not move steadily forward from success to success. People battered their heads against the barrier of the τ/θ puzzle for years until Lee and Yang showed that the door opened in an unexpected direction. 'Great ideas' (analyticity, Reggeism) proved only partial illuminators

158

of murky scenes. Ideas developed for one purpose turned out to find their most promising applications in other ways (duality and strings). Notions of great eventual fruitfulness (electroweak unification and intermediate vector bosons) were for long thought of as the preserve of rash and eccentric speculators. New quantum numbers were invented (strangeness, charm) and even the notion of a constituent particle had to be revised (*confined* quarks). Throughout the period under review there was an interplay of experiment and theory much more subtle in its character than can be expressed by the classic oversimplification of the unproblematic confrontation of prediction by measurement. The intransigence of τ/θ drove theorists against their will to acknowledge a handedness in weak interactions; the ultimate recognition of the attractiveness of electroweak theory drove experimentalists to reconsider their background calculations and to renew the search for neutral currents.

In the short term (from Conference to Conference) it is a lurching story of both bafflement and breakthrough, with fluctuating fortunes (the ups and downs of the $|\Delta I| = \frac{1}{2}$ rule), pseudoproblems (pion beta decay), errors perpetrated (the split A_2), errors corrected ($V - A$), triumphant vindications (parity non-conservation, two neutrinos), and unexpected discoveries (*CP* violation, J/ψ). In the longer term (the whole period under review), when there is time for an averaging out of experimental vagaries and a critical sifting of theoretical speculations, the character of the story is clearer. In 1950 the high-energy physics community believed protons and neutrons to be fundamental particles, participating in a physical structure characterised by four forces of nature, one of which (the strong nuclear force) was of a largely unknown character. In 1980, the high-energy physics community believed protons and neutrons to be composites composed of quarks and gluons, participating in a physical structure which included as a component a unified electroweak force, which in turn might prove capable of further unification with the strong forces, the latter being in any case described by a specific gauge theory, quantum chromodynamics. Whatever arguments and dissensions there might have been on the way to this conclusion, by 1980 virtually all high-energy physicists subscribed to such an account of the pattern and structure of the physical world. It was the Standard Model. In a word, the dust had settled. How that was achieved, and what was the actual nature of that achievement, are questions which the philosophy of science seeks to address.

Consider the evolution of the concept of quarks. At first, the suggestion is made as a way of thinking about properties of resonances (the quark model) or currents (current algebra). At this stage quarks are a heuristic device, a fruitful manner of speaking which leads to observable relationships but which does not demand a realistic interpretation for these 'presumably mathematical' entities. Indeed, there are grave difficulties for such an interpretation — the lack of observation of fractionally charged particles and the symmetrical character of the baryon ground state (the 56). Both these problems prove to be susceptible to what are deemed to be natural explanations: the idea of confinement, though not demonstrated, is at least motivated by the thought that infrared slavery could be the obverse of asymptotic freedom; the colour degrees of freedom provide the antisymmetry necessary for the 56; whilst quark-counting experiments (π° decay, e^+e^- annihilation) are strongly indicative of just such a tripling of quark states. The eventual universal acceptance of quark theory results from further developments, both experimental and theoretical. Deep inelastic scattering experiments, in both electroweak and strong interactions, are readily interpretable as evidence for the presence of pointlike constituents with quark quantum numbers. Quarks and leptons are incorporated into Salam—Weinberg electroweak unification, and quarks and gluons provide the basis for an elegant theory of strong interactions (QCD).

It cannot be said that the standard model is _entailed_ by the data, but despite the restless ingenuity of theorists, ever seeking reputation through the suggestion of alternative ideas, it is the only theory available which is fully in accord with general principles (relativistic quantum mechanics), is economic and elegant in its construction (free from an air of undue contrivance), and is in general agreement with a vast range of observations of diverse kinds. That agreement cannot be said to be perfect — there are great difficulties in eliciting reliable estimates of scalebreaking effects in deep inelastic scattering, for instance — but it appears adequate within the limitations imposed by theoretical and experimental uncertainties.

A thoughtful physicist asked about method would be unlikely to deny the role of judgement in the evaluation of what is happening. Experiments are never perfect, background estimates depend upon a critical identification and assessment of possible competing effects. Some predictions of theory are sharp (the masses of the intermediate vector bosons follow from Salam—Weinberg theory once the value of θ_w has been established), others are distinctly fuzzy (scalebreaking corrections), yet others depend upon questionable model assumptions (attempts, bag model or otherwise, to calculate the hadronic spectrum in QCD). It is the essence of science to interrogate nature from a point of view, just as it is the essence of science to be open to the correction of that point of view. Experiments are certainly theory-laden. The τ/θ decays seemed to show a mysterious degeneracy of particles of opposite parity to those who believed that parity was conserved; to Lee and Yang they just indicated the violation of parity. Deep inelastic scattering needed a parton interpretation before it was seen as pointing to constituent quarks. Yet theories are also experimentally sifted. Parity non-conservation led to a number of immediately testable predictions. Without these having been verified the idea would not have been accepted. The quark parton model did not lend itself to such absolutely clear testing, but analysis of the structure of jets found in processes with large amounts of momentum transfer is an important encouragement to its acceptance.

The same thoughtful physicist asked about achievement would say that we have discovered a new level in the structure of matter. We now know that protons and neutrons are made out of quarks and gluons. He would not just be saying that this was a useful way of thinking about things. He would believe that he was talking about what is actually the case. Philosophically, physicists are realists. To them the history of high-energy physics in the period 1950−80 is that of a tightening grasp of an actual reality. We have found out something about the constitution of the physical world that we did not know before.

Most physicists are bewildered by having any other proposition put to them. How otherwise could one explain the success of science? Indeed, what other understanding would make the whole enterprise worthwhile? As the philosopher Hilary Putnam is fond of saying, to realists it would seem a 'miracle'[4] that talk of electrons and quarks and all that should provide so good an account of phenomena if there were not actually electrons and quarks and all that lying behind the phenomena. 'The positive argument for realism is that it is the only philosophy which does not make the success of science a miracle.'[5] Yet very substantial criticisms have been made of the thesis of scientific realism. We must consider some of the issues which have been raised.

1   It is sometimes claimed that modern physics, and in particular the elusive and counterintuitive character of quantum theory, has killed realism. 'Realism is dead ... Its death was certified finally as the last two generations of scientists turned their backs on realism.'[6] People who take this point of view often believe Niels Bohr to have administered the fatal dose.

It would be as narrowly parochial to take the commonsense of the everyday macroscopic world as the touchstone of reality as it would to offer a similarly privileged position to information derived from electromagnetic radiation of wavelengths of a few thousand angstroms (the visible spectrum). One should not commit the error of equating reality with ordinary objectivity. That was the mistake Einstein made and which then drove him to oppose the modern version of quantum theory. The reality of the quantum world (which I have sought to defend elsewhere[7]) is to be construed in accordance with its own idiosyncratic nature. It is my contention that ultimately it is the intelligibility of the quantum world which is the proper ground for belief in its reality. Certainly, reality is not to be restricted merely to the picturable.

161

Little billiard balls are not the only conceivable form of constituents, and if we can make sense of matter's properties in terms of a more subtle structure, then we should take that structure with the utmost seriousness. As for Bohr, it is notoriously difficult to pin down his philosophical position — he was an extensive but cloudy writer on such matters — but commentators find a realist intent in what he was saying.[8] I claim that the self-understanding of the high-energy physics community in the period under review is a direct denial of Fine's assertion that the physicists have abandoned realism. The motivation of almost all that community's members was to find out what the constituents of matter actually were and how they actually behaved.

2   It is frequently claimed that there are historical discontinuities in the development of science which destroy the possibility of its being concerned with the discovery of truth. In 1950 protons and neutrons are elementary; in 1980 they are composite. In 1980 quarks are elementary; in 2000 will they be excitations of strings or whatever? Who can place confidence in so shifting an account? Newton-Smith has formulated the 'pessimistic induction': 'any theory will be discovered to be false within, say, 200 years of being propounded'.[9]

Certainly, this observation is destructive of any naive claim that science establishes absolute truth. Its conclusions are always provisional and corrigible in the light of further exploration. But those of us who want to defend a critical version of scientific realism would want to assert that science does achieve verisimilitude, 'an adequate account of a circumscribed physical domain, a map good enough for some, but not for all, purposes'.[10] The claim is that the development of a mature, well-winnowed scientific theory is comparable to our inspection of a figure approaching us across the moors: a dark blob — a matchstick man — a person in a kilt — Fergus. The description changes, but that is only due to a finer-grained perception of the same entity.

This continuity of reference with change of account is denied by Thomas Kuhn and his followers. They assert that there are revolutionary periods in science where the world-view (or paradigm, as they say) changes so as to produce a radical disjunction between the old science and the new. In a famous passage, Kuhn claimed, 'In a sense that I am unable to explicate further, the proponents of competing paradigms practise their trades in different worlds.'[11] This is the celebrated Kuhnian thesis of incommensurability. According to this view there is no dictionary which would enable Newton and Einstein to speak to each other. In the history of physics there is J.J. Thomson's electron, Bohr's electron, the electron of modern quantum electrodynamics — all strictly incomparable entities. Where there is no basis for comparison, there is also no rational ground for discriminating theory choice nor a sustainable account of scientific progress. A strictly Kuhnian view of the history of physics would be the story of competing ideologies and their respective propaganda machines. In his later writings Kuhn has recoiled somewhat from this drastic claim.

So startling and implausible a conclusion only follows if one waywardly insists on defining entities holistically, that is to say, in terms of the totality of all their properties. It then becomes a matter of a package deal. Either you believe all that J.J. believed about the electron or you are not talking about what he was talking about. Frankly, that's absurd. One needs a certain 'charitability of reference'[12] which takes account of the advance of knowledge, so that Thomson is construed as wishing to refer to the same electron that we now describe by the greatly refined theory of quantum electrodynamics. It is not hard to find partial but continuing characteristics of electrons which can be used to explain how to exercise such charity: emitted from hot cathodes, mass of about 10^{-27} grams, charge of about 5×10^{-10} esu, etc.

Andrew Pickering has claimed, nevertheless, to be able to discern a revolutionary paradigm shift within high-energy physics during the very period that we are surveying. He defines an 'old physics', characterised as being concerned with resonances and scattering at low momentum transfer and using Reggeism as a principal theoretical tool, and a 'new physics', characterised as being concerned with quarks and scattering at large momentum transfer and using the standard model of electroweak−QCD as its principal theoretical tool. He asserts that, 'The old and the new physics constituted, in Kuhn's sense, distinct and disjoint worlds.'[13] I disagree.

There was, of course, a shift of interest and effort between the 1960s and the 1970s which moved from concentration on the 'old' topics to concentration on the 'new'. The reason lay precisely in that 'opportunism in context' which characterises the exploration of the physical world. It was realised that deep inelastic scattering provided that 'simplicity through extremity' which is always needed to get some purchase on the structure underlying the diversity of phenomena. The construction of the standard model fully justified that strategy. In terms of that model the old problems in no way became unstateable or meaningless. They were simply recognised as being intrinsically more difficult than the 'easy' cases at large p_T. The various bag models, together with lattice gauge theory, represent attempts to unravel something of the standard model's implications for resonance physics. 'What are the Regge properties of QCD?' is a perfectly coherent question to pose. An attempt to answer it will encounter complexities analogous to those of the Reggeon calculus and, in my opinion, very similar techniques will be needed in both cases. Gell-Mann said recently of Regge theory that 'These

163

singularities are just as important as ever in spite of the fact that they are a less fashionable research topic than they once were.'[14] The strong bandwagon effect in high-energy physics, which encourages people to think that only the current vogue issue is worth discussing, certainly discourages theorists from undertaking the formidable task of elucidating Reggeism in QCD, but it is fashion and difficulty, not incommensurability, which holds them back. It is just not true to say that 'the new-physics theories had nothing to say on the common phenomena of the old physics'.[15]

A much sharper threat of discontinuity occurs when the change is not in the tone of discourse but instead it turns out that what was thought to be a well-established _entity_ disappears from that discourse altogether. The 'tightening grasp of reality' then turns out to be the the closing of an empty fist. Larry Laudan has made much of this as a refutation of the sort of convergent realism that I am seeking to defend. With italic emphasis he tells us that 'a realist would never want to say that a theory was approximately true if its central terms failed to refer'[16] and he goes on to give a long list of just such failures. It is perhaps significant that none is later in date than the turn of the nineteenth century. I cannot think of an example in twentieth-century particle physics.[17] Let us consider what is the latest and the most troubling of Laudan's examples, the disappearence at about the beginning of this century of the aether from the list of physical entities. Maxwell had described this subtle medium, believed to be the carrier of electromagnetic waves, as better confirmed than any other entity in natural philosophy.[18] Now it is no more. Michelson and Morley and Einstein have removed it from the portfolio of physics.

I would want to make two points. The first is that although the aetherial Cheshire cat faded away, its grin did remain. I do not mean simply the structural continuity of Maxwell's equations (first thought to describe aether vibrations but still remaining fundamental to the description of the electromagnetic field) but also that the aether was a mechanical way of thinking about the enduring concept of a field and that the quantum-mechanical vacuum is also an all-pervading medium of subtle property in which excitations of energy (particle blips) occur. There is a sense in which the aether was a mistaken way of looking in a fruitful direction, not altogether dissimilar to the way in which the thoughts of the early atomists held some promise, even if the constituents of matter are not hard spheres with hooks. The other point is that, however blurred physics' early vision may have been, increasing perspicuity has not in fact resulted in the discovery of a sequence of dissolving mirages. In assessing the significance of the fact that the fundamental physics of the twentieth century has not had to discard established entities, one must bear in mind that the generations of knowledge have become shorter with increasing rapidity of exploration. It took a hundred years to establish modern atomism. In the last eight decades we have seen three further generations, moving from atoms to nuclei, then to hadrons and then on to quarks. The length of our experience of change in fundamental physics is not measured in years alone. We therefore have good inductive grounds for convergent realism. Of course we do not have certainty (a point to which I shall return). Hilary Putnam may at times feel haunted by the nightmare, 'What if all the theoretical entities postulated by our generation (molecules, genes, etc., as well as electrons) invariably do not exist from the standpoint of later science?'[19] The snark *might* turn out to be a boojum, but it does not seem in the least degree likely.

3   I am defending an account of science which claims a convergent realism to be the nature of its achievement. In Newton-Smith's words the assertion is of a *'thesis of verisimilitude . . .* the historically generated sequence of theories of a mature science is a sequence of theories which are improving in regard of how approximately true they are.'[20] Philosophers have challenged the coherence of such an assertion. How can one measure the degree of one's proximity to reality? Pontius Pilate's question is to be amended to the inquiry, 'What is approximate truth?' There are two issues: what could such a claim mean and how would we know if it were true?

I do not think that the first problem is to be solved simply by seeking to establish some measure of the quantity of true and false statements entailed by a theory and then requiring the truth/falsehood ratio to increase with increasing verisimilitude.[21] Formal arguments like these bear little relation to scientific experience. They fail to incorporate any discrimination between significant and trivial statements and without such a weighting the comparisons degenerate into meaningless ratios of infinities. Once a theory has some margin of error it necessarily entails an unbounded quantity of falsehood, but perhaps of a recognisably restricted kind. Given special relativity, Newtonian mechanics is to some tiny amount always wrong, and it is wrong to a substantial amount for particles whose velocities are an appreciable fraction of the velocity of light. The sensible way to think of this is not just to count errors but to recognise the highly verisimilitudinous character of Newtonian mechanics in extensive, but circumscribed, domains of physics. (It is good enough to use to send a space-probe to Mars.) I believe that judgements of verisimilitude require tacit skills, not specifiable in advance according to some success-counting algorithm, but exercisable according to recognisably rational principles. A key role is played by the correspondence principles of physics, which indicate why the old theory and the new coincide asymptotically in the bounded domain of the former's successful application. Such correspondences involve not only close coincidence of numerical predictions, but also conceptual correlations, so that we understand how the entities of quantum theory are reconciled with J.J. Thomson's belief that he had discovered particulate electrons and Thomas Young's belief that he had demonstrated the wavelike character of light. Such 'observational nesting' (as Newton-Smith has called it[22]) is analogous to our experience that the kilted figure, spotted when five hundred yards away, can be discriminated and recognised as being Fergus and not Alexander, when he is only one hundred yards away.

It is the experience of such historical sequences of theories, linked by correspondence principles and capable of being regarded as successive refinements of each other, which is the basis of the claim to be able to recognise increasing scientific verisimilitude. I am very sympathetic to Ernan McMullin's statement that 'realism . . . is, in part, an empirical thesis'.[23] We believe it because it has been found to work. 'It is, in part at least, because the history of science testifies to substantial continuity in theoretical structures that we are led in the direction of scientific realism at all.'[24] In order to make such an assessment one must be prepared to allow time for a winnowing process to take place. The dust must settle. In the shorter term, ultimately accepted theories may not score much initial success. Think of the early days of theories of weak interactions using intermediate vector bosons. Yet today we know that the W and the Z exist and we also understand why the Fermi theory works well at moderate energies. Equally, in the shorter term, ideas may score some successes without anyone expecting them to be enduring parts of fundamental physics (for example, bag models). It is helpful to discriminate between theories (which are bids for achieving a fundamental basis for understanding what is going on) and mere models (which are frank attempts simply to get some purchase on a set of phenomena, in terms which are helpfully suggestive rather than ontologically serious in detail). No one believes that hadrons are really bags of quarks (even in the most relativistically sophisticated sense of 'bag'), but quantum chromodynamics is a serious candidate for providing a fundamental, widely applicable, understanding of quark dynamics.

One of the strongest arguments for verisimilitude is when concepts prove to have a fruitfulness going far beyond the phenomena which first gave rise to their suggestion. Intermediate vector bosons enabled the successful unification of weak and electromagnetic interactions to be achieved but the first reason for thinking about them lay simply in a desire to make $(V-A)$ look natural. CP-violation offers an explanation of why the universe is matter rather than antimatter. Such successes do indeed seem 'miraculous' if there is no grain of truth in the original idea.

4 If it were not the way things are which gives rise to scientific theories, then what could it be that does so? It is frequently claimed that the

sociology of knowledge provides the answer. It is the scientists who make the science. By the largely unconscious adoption of certain modes of experimental interpretation, and by compliance with culturally conditioned theoretical expectations, it is asserted that an order is imposed upon the physical world, not read out from it. Barry Barnes claims that:

> Progressive realism is one of the ideal accounts of scientific knowledge which has it moving towards something, in this case a description of the real existing mechanisms of the world. There are now several independent strands of work which imply that such theories are misconceived, and that all knowledge generation and cultural growth should be regarded as endlessly dynamic and susceptible to alteration just as is human activity itself, with every actual change or advance a matter of agreement and not necessity.

In his view, 'our present theories stand symmetrically with earlier scientific theories, and for that matter with any other'.[25] Our tale would then indeed have been one of a roundabout, a ceaseless circling which never got anywhere.

In his book *Constructing Quarks*, Andrew Pickering has argued just such a thesis in relation to those developments in elementary particle physics which led to the standard model. In his opinion, 'The world of H[igh] E[nergy] P[hysics] was *socially* produced.'[26] 'The quark-gauge theory picture of elementary particles should be seen as a culturally specific product.'[27]

What arguments are produced to support this astonishing conclusion? One is the claim of the incommensurability of the old and new physics. If that were true then there was a theoretical lurch from one world to another, so that 'The particle physicists of the late 1970s were themselves quite happy to abandon most of the phenomenal world and much of the explanatory framework which they had constructed in the previous decade.'[28] I have already given reasons for rejecting incommensurability.

A second line of argument arises from the analysis of certain episodes in the story. In the 1960s bubble-chamber physicists thought that their background calculations excluded the then theoretically unwanted possibility of neutral currents. In the 1970s, when neutral currents came into theoretical favour, they revised their calculations, did further experiments and concluded that the effect was present. This is certainly an instructive episode. Experimentalists do find it difficult to see what they are not expecting. Background calculations call for tricky acts of judgement and they are more likely to be accepted with relief when they result in conclusions in agreement with current prejudice. One must acknowledge that social influences play a part in scientific investigation − both in determining what is considered to be significant and worth the toil of testing, and in producing a frame of mind about what is likely to be found. Yet one must also acknowledge that the

167

stubborn facticity of nature imposes ineluctable constraint, whatever one might have anticipated would be the case. Discoveries such as *CP* nonconservation and the J/ψ were totally contrary to prior expectation. I deny that neutral currents are just an artefact of how the scientists decided to analyse the experiments. There are entirely rational grounds for accepting the later background calculations to be the better based. A leading experimentalist engaged in the experiments of both the 1960s and of the 1970s agreed in conversation that this was so. He lamented the fact that they had missed the earlier chance of a sensational discovery by complacently accepting the original analysis. The instrumentations had also become more sophisticated and discriminating in the 1970s. I therefore cannot agree with Pickering when he writes of 'the different interpretative practices in neutrino physics which determined the existence or nonexistence of the neutral currents' that the 'choice was irreducible; it cannot be explained by comparison between predictions and data which were its consequence'.[29]

   A similar point can be made in relation to another example discussed by Pickering. A consequence of the existence of weak neutral currents is that one expects some minuscule degree of parity nonconservation in the interaction of electrons with matter. Its investigation is an exceedingly delicate question because the effects will largely be masked by the much greater parity-conserving effect of electromagnetism. Two ways of investigating the matter were tried. One involved measuring properties of atomic levels. Original experiments at Seattle and at Oxford failed to reveal the calculated expectation. Another experiment at SLAC, brilliantly skilful in its execution, looked directly for parity-violation in electron−proton scattering. It agreed with the Salam−Weinberg predictions. Pickering presents the acceptance of the latter and the rejection of the former as pointing to 'the local construction of a self-consistent world of harmonious phenomena and theory',[30] as if this were a matter of whimsical choice. Such a comment is tendentious in the extreme. It fails to recognise the doubtful character of the atomic level calculations needed by the Seattle and Oxford groups and the clean character of the SLAC experiment, whose interpretation was free from these additional complications. There are entirely rational grounds for preferring the conclusions of one experiment to those of the others.

   5   The idea that the order science finds is an order that it imposes − that we construct the pattern of the physical world rather than discovering it − is not a product solely of the modern sociology of knowledge. Its history stretches back at least as far as the writings of Immanuel Kant at the end of the eighteenth century. Kant believed, for example, that three-dimensional Euclidean geometry was an a priori notion, necessary for our being able to come to terms with the flux of experience. One would be more impressed

with the quality of the argument if it had led him to recognise that Euclidean geometry was just one of a number of empirically open possibilities! Whatever forms such claims take they depend in their bid for plausibility upon there being a certain degree of plasticity of physical theories in relation to phenomena, a plasticity which is asserted to permit us to bend the account to the desire of our pattern-seeking will. Such flexibility, if it were actually present, would certainly put in question a realist philosophy of science.

One of the commonest arguments invoked by antirealists is that of the underdetermination of theory by data. The results of experiments, however numerous, only furnish us with a finite number of measurements. The claims of theory purport to cover an infinite number of instances. There is a perpetual mismatch between what is ascertained and what is claimed to be known. The other side of that coin is the statement that there are unlimited theories capable of 'explaining' limited sets of facts.

Philosophers of science who argue in this formal vein frequently remind one of the constructive quantum field theorists (p. 138). They are so obsessed with what is logically demonstrable that they fail to recognise what is scientifically interesting. One feels that the paradigm they have in mind for theory construction is that provided by arbitrary curve fitting. Given a million and one points, there are an infinite number of curves (polynomials of degree greater than or equal to a million) which will pass through them. So what? If science proceeded in this way it would have no long-term fruitfulness nor its own intrinsic rationality. Each new measurement would require a revision of the curve-fitting procedure, just as Ptolemy could only have maintained his theory post-Newton by adding further epicycles each time astronomical observations improved in accuracy. Temporary empirical adequacy would be the only operative criterion. Such a picture is a travesty of physics.

Scientific theory is in fact constrained by two sets of criteria, both of which will call for further discussion later on in the argument. One set refers to rational evaluation based on such general qualities as simplicity and fruitfulness. The other set refers to general physical principles (currently those provided by relativity and quantum theory) which are not treated as being absolute and incorrigible but which, on the other hand, have played so significant a part in scientific understanding over so long a period that they are by no means lightly to be abandoned. It is in fact a highly non-trivial exercise to find a theory of empirical adequacy which satisfies these criteria. Philosophers of science seldom exhibit any understanding of how difficult it is to produce even a passable shot at a credible explanation of a wide range of phenomena. The whole burden of our tale has been of the search for a theory capable of providing an understanding of the discoveries of postwar high-energy physics. After much struggle, the standard model emerged as the sole available contender. Given the restless and competitive spirits of bright young theorists, ever anxious to gain reputation by refuting or replacing the ideas of their elders, it is hard to believe that there are many, or any, rational alternatives which have escaped notice through a socially induced slothful acquiescence in the status quo. (Think of those sixty or so papers at London 17, all failing satisfactorily to explain the rise in R in e^+e^- annihilation.) Of course, the research has been conducted within the general rational constraints described above, but there seems no reason whatsoever to follow the spritely but implausible advice of anarchic irrationalists like Paul Feyerabend[31] and to suppose that astrology or witchcraft would have provided alternative intellectual perspectives of equal fruitfulness. In the actual scientific enterprise, the problem is not the embarrassing riches of a superfluity of theories but the extraordinary difficulty in finding one that is at all believable.

Each of the critical attacks on scientific realism contains an element of truth: (1) the physical world is not naively objective in an everyday sense but it corresponds to a more subtle kind of reality; (2) physical theory is always corrigible. The exploration of new regimes often reveals unexpected new features, without destroying the circumscribed but actual value of the theory previously successful; (3) attempts to make more precise the notion of verisimilitude are certainly to be encouraged, but they will require taking into careful account the actual character of scientific inquiry; (4) science is a social activity, with the dangers of fashion and tricks of perspective which follow from that, but its conclusions are not to be treated as simply a socially determined consensus; (5) physical theory is not just read off from data but it is constructed according to rational criteria, involving the exercise of tacit skills rather than automatic analysis.

These facts explain why scientific realism has to be a critical realism. They do not destroy the claim, almost universally made by scientists themselves, that they are investigating a physical world which in its idiosyncratic reality controls and determines our knowledge of it. They do mean that science bears a kinship to many other forms of human inquiry. It is not wholly different from, and superior to, other forms of rational discourse, but rather it takes an honoured place among them.

It is instructive to turn to the thinking of those philosophers who have sought to give a rational account of scientific method and its achievement. Foremost in the minds of many people would be Karl Popper. The minority of scientists given to adopting an explicit philosophical point of view often declare themselves to be Popperians. I think that they do so for two reasons. First, Popper's heart seems to be in the right place. He says things like, 'Our main concern in science and philosophy is, or ought to be, the search for truth.'[32] But with Popper, who in his essential being is a logician, it is the head that rules the heart, with the result that a narrow reliance on deduction creeps into his formal account of scientific method. The second reason for Popper's popularity among scientists is that his emphasis on the role of falsification appears to reproduce a recognisably significant aspect of the way they work. Physical theory is continually being forced to place its head upon the experimental chopping block. Bell and Perring have an ingenious idea about the source of *CP* non-conservation. It implies an effect increasing with the square of the energy. Does this happen? It does not — so curtains for Bell and Perring. Thus the Popperian account of 'bold conjectures' and 'the critical search for what is false in our competing theories'[33] does have a specious plausibility as the story of what is happening.

It will not do, however. It fails in various respects. The emphasis on bold conjectures makes science sound like a sort of intellectual shooting gallery in which the scientist takes a succession of pot shots. It is a fact of experience that fruitful conjectures are not simply bold but they are constrained by the sort of rational criteria we have been outlining. The emphasis on refutation — 'Only the falsehood of a theory can be inferred from empirical evidence, and this inference is a purely deductive one',[34] a chilling message conveyed in italics in the original — gives a curiously cockeyed view of the scientific endeavour. When Rubbia and his friends rejoiced at the outcome of the UA1 experiment and believed that they had discovered the W and Z bosons, on a Popperian account their happiness was misplaced. Truth is always unknowable, the only certain knowledge is that of error. On that view, what would really have justified a party would have been the failure to discover the predicted W and Z signals! There is something very awry in such an account.

When Popper's heart has temporarily taken over from his head, he sees that this is so. In reply to critics he wrote,

> ... it would be a highly improbable coincidence if a theory like Einstein's could correctly predict very precise measurements not predicted by its predecessors unless there is 'some truth' in it ... 'a higher degree of verisimilitude *than those of its competitors*' ... there may be a whiff of inductivism here.[35]

Bill Newton-Smith comments that there is not a whiff of inductivism here but 'a full-blown storm'.[36] The truth of the matter is that no account of science will be adequate which does not grasp the nettle of induction and so does justice to the experience of gaining actual knowledge (we have learnt that there are Ws and Zs). The further truth of the matter is that induction cannot be justified on narrowly logical grounds. There is no algorithm specifying how many 'for instances' are necessary before we have grounds for belief. Such judgements are, once again, to be made by the exercise of tacit skill and submitted to a competent community.

A further difficulty for Popper is to give a clear account, in realistic circumstances, of when refutation takes place. Very general and powerful principles are not to be discarded at the first suggestion of an adverse result. In 1921, when Einstein was told that Miller had measured a non-zero aether drift, he did not feel that special relativity was in danger but he serenely replied, 'Subtle is the Lord, but malicious he is not.'[37] Imre Lakatos has tried to take this into account in his description of scientific research programmes.[38] These contain a hard core of central ideas, non-negotiable as far as that programme is concerned, surrounded by auxiliary hypotheses which can be adapted to a greater or lesser extent to cope with threatened anomalies. That is certainly a step in the right direction of a realistically recognisable account of scientific activity. Relativistic quantum mechanics constituted the hard core of high-energy physics during the whole period under review. It would only have been abandoned under the most severe and sustained pressure, and its modification would have constituted a profound alteration of physical theory. More specific and detailed ideas were open to revision. Some were judged to be of sufficient promise to be worth holding on to at the cost of a degree of ingenious modification unforseen at their inception (colour and confinement were introduced into quark theory); others were so peripheral that their demise caused no stir (the dubnon).

The problem of justifying induction has caused some philosophers to retreat from pushing ontological claims (of a knowledge of what is) in favour of science's simply achieving instrumental ends (the ability to get things done). Richard Rorty takes the pragmatic view that 'modern science does not help us to cope because it corresponds, it just plain enables us to cope'.[39] But if science enables us to cope, how does it possess this miraculous property except by some convergent correspondence to the way things are? It surely does not arise as the spin-off from a socially agreed construction. 'It cannot be the explanation for the fact that airplanes, whose design rests upon enormously sophisticated theory, do not crash is that the _paradigm_ defines the concept of an airplane in terms of crash resistance.'[40] And the enduring fruitfulness of central scientific concepts is not to be explained away in a positivistic sense by their being just convenient summaries of data. The successes of the quark model in giving understanding of behaviours in deep inelastic domains goes way beyond any minimal role of a convenient mnemonic for patterns of resonances.

172

Ian Hacking has written extensively on instrumental success as the endorser of scientific realism.

> Experimental physics provides the strongest evidence for scientific realism. Entities that in principle cannot be observed are regularly manipulated to produce new phenomena and to investigate other aspects of nature. They are tools, instruments not for thinking but for doing.[41]

Hacking believes in electrons because he believes in the electron microscope. Yet this leads to an etiolated form of realism. Hacking thinks that, 'The vast majority of experimental physical physicists are realists about entities and not about theories.'[42] He goes on to say, 'My attack on scientific antirealists is analogous to Marx's onslaught on the idealism of his day. Both say that the point is not to understand the world but to change it.'[43] That will not do. Scientists are just not satisfied with instrumental success alone. They demand understanding as their ultimate goal. The $|\Delta I| = \frac{1}{2}$ rule has had good predictive success but it vexes physicists that they do not have a theory to explain its origin. No utilitarian purpose drove the search for the W and the Z but rather a desire to confirm the beautiful rational structure of electroweak unification. Current speculations about superstrings could only have effects in regimes many orders of magnitude beyond conceivable accessibility, but they are pursued because, if true, they would illuminate the question of the deep structure of matter.

I believe that the use of inductive argument is fundamental to the scientific endeavour, yet it eludes capture in any algorithmic net. Science is an activity of judgement, involving tacit skills. In Polanyi's oft-repeated phrase, 'we know more than we can tell'. Science is a convivial activity, pursued in a community to whose judgement the individual scientist offers his efforts as a corrective to merely individual idiosyncracy, but controlled and determined by the stubborn facticity of the physical world whose understanding is being sought. Not detachment but intellectual passion is involved, for 'the discovery of objective truth consists in the apprehension which commands our respect and arouses our contemplative imagination'.[44] Hence the use of 'wonder' as a word to describe the scientist's satisfaction in his work.

Scientific method can be described in action and learnt through its experience − it is a craft to which one must serve an apprenticeship − but it is not exhaustibly characterisable in prior logical terms. There is no philosophical programme which will reveal 'a methodologist's stone capable of turning the dross of the laboratory into the gold of theoretical truth'.[45] Yet one can recognise rational characteristics which are to be demanded of a scientific theory if it is to be accorded the epithet 'good'. Kuhn suggested five: (i) accuracy; (ii) consistency; (iii) wide scope; (iv) simplicity; and (v)

173

fruitfulness.[46] Clearly a criterion such as simplicity involves in its application an act of personal judgement. It is extremely interesting that such a test (particularly in the form of the economy and elegance which we call the mathematical beauty of a theory's formulation) has time and again proved consistent with the criterion of fruitfulness. Physicists habitually use simplicity as a heuristic device in the search for fertility. Dirac once said of himself and Schrödinger that, during their fundamental work developing modern quantum theory,

> It was a sort of act of faith with us that any equations which describe fundamental laws of Nature must have great mathematical beauty in them. It was a very profitable religion to hold and can be considered as the basis of much of our success.[47]

I cannot accept Kuhn's claim that such criteria are of merely persuasive, rather than evidential, character.[48] Instead I wish to adopt the stance, called by Newton-Smith 'temperate rationalism',[49] which sees science as the reasoned pursuit of verisimilitude, based on a style of argument which is often summarised as 'inference to the best explanation'. It is a total explanatory framework which is being sought, not just the limited objective of 'inference to the most likely cause' of particular phenomena, as Cartwright has suggested.[50] The explanation thus achieved is to be taken with ontological seriousness. Its entities are not just means to the end of an empirical adequacy, a mere 'saving of the phenomena' (accounting for observations), as the constructive empiricism of van Fraassen alleges.[51]

Abduction (as inference to the best explanation is called) has been challenged by Laudan to justify itself by its own method. 'If realism has made some novel predictions or has been subjected to carefully controlled tests, one does not learn about it from the literature of contemporary realism.'[52] The just comparison would be with observational sciences, such as cosmology or evolutionary biology, with their power to illuminate historical process. A cogent defence of realism can attempt to

> argue from progress to the viability of a methodology. Scientists in choosing between theories do not act capriciously. They deliberate and in the dialectical process of discussion they offer reasons for their choice. Given that there has been progress, we have reason to think that the procedures they follow are by and large evidential. That is, in general at least, the considerations that motivate them in selecting theories are fallible indicators of verisimilitude.[53]

The words are Newton-Smith's. He offers his own list of criteria for 'good-making features of theories': (i) observational nesting (p. 166); (ii) fertility; (iii) track record; (iv) inter-theory support [but remember the present imperfect reconciliation of quantum theory and general relativity]; (v) smoothness; (vi) internal consistency; (vii) compatibility with well-grounded metaphysical beliefs [such as causality]; and (viii) simplicity.[54] By 'smoothness' is meant a theory's capacity to accommodate increasingly accurate and testing experimental probing by adjustments which lack contrivance and which arise naturally from within the theoretical point of view adopted. The successive refinements of quantum electrodynamics constitute an outstanding example of the achievement of such smoothness.

I think that the most satisfying account of a critically realist and temperately rational point of view has been given by someone who was himself a distinguished practising scientist, Michael Polanyi.[55] He believed the true character of scientific knowledge not to be objective but to be personal, pursued individually but within a convivial community and with universal intent. The epithet 'personal' indicates the character of the inquiry (it could not be conducted by a computer, however cleverly programmed, for it involves the exercise of tacit skills of judgement) but it does *not* imply an idiosyncratic outcome. The scientific community constitutes a 'competent authority' but the 'supreme authority' is physical reality itself.[56] Polanyi has been strangely neglected by philosophers of science (even Newton-Smith, whose stance seems so close to Polanyi's, does not refer seriously to him) but I think his account is broadly persuasive just because he speaks from prolonged experience of how science is actually conducted.

We possess the rational skills which enable us to investigate the nature of physical reality. The scientific world view is always provisional rather than final, its achievement verisimilitude rather than absolute truth, but it offers us an ever more adequate understanding of the pattern and structure of the physical world in which we participate. That is what was happening in the story of high-energy physics.

It is a very remarkable property of that world that it is open to our inquiry in this way; that the rationality we experience within (our mathematically based theories) and the rationality we observe without (the experimentally apprehended way things are) fit together so perfectly. I think that it is perverse of Pickering to claim that 'given their extensive training in mathematical techniques, the preponderance of mathematics in particle physicists' accounts of reality is no more hard to explain than the fondness of ethnic groups for their native language'.[57] In actual fact, the situation is precisely the reverse. Physicists laboriously acquire the language skills of mathematics because their use has proved to be the uniquely fruitful way by which to understand the physical world. Nor will the need to survive in the evolutionary struggle provide the necessary link between the reason within and the reason without. To be sure, our thoughts and experience must match

175

at the everyday level if we are not to injure ourselves or die of starvation, but that fails totally to explain why the abstract mathematics of gauge field theories fits so perfectly the world of high-energy physics, so utterly remote from the perceptions which relate to macroscopic phenomena. The universe is marvellously rationally transparent to our inquiry. The instinct of a scientist to seek an explanation through and through will not allow him simply to say, 'that's just the way it is, and good luck for those of us who are mathematically able.' He will want to go beyond his science (with its _assumption_ of cosmic intelligibility) to a wider setting in which that intelligibility will find its own explanation. In my view, that search will take him in the direction of theology. But that is another story.

Appendices

These seek to treat a few topics in a little more detail than is appropriate in the main text.

Appendix 1   Regge Theory

An important concept is that of signature. It arises from the existence of crossed channels. A fully relativistic account is available but the essential idea is most easily explained in terms of a potential theory model. The analogy to that interchange of particles which crossing implies is then the existence of exchange potentials (V_1) by means of which particles not only interact but are also interchanged. (Such potentials have long played a role in nuclear calculations.) In a state of angular momentum l the interchange produces a factor of $(-1)^l$. Thus, if there is also an ordinary potential (V), the total effect is $V + V_1$ in even angular-momentum states and is $V - V_1$ in odd angular-momentum states. This implies that there must be distinct families of Regge poles for the even and odd cases. They are said to be of even or odd signature, respectively. The even- (odd-) signature Regge poles only give particle poles at even (odd) physical values of angular momentum. However, both sets of poles contribute to the high-energy behaviour.

The optical theorem relates the total cross-section to s^{-1} times the imaginary part of the forward scattering amplitude. If the cross-sections are to be constant at large s, then the forward amplitude will have to contribute a factor of s to cancel the kinematic factor of s^{-1}. This will be achieved if there is a Regge pole (the Pomeron) with vacuum quantum numbers (so that it can be exchanged in any process), passing through 1 at $t = 0$ (the forward direction), and of even signature (so that it does not give unwanted particle pole at $t = 0$).

177

Appendix 2   Sum Rules

I use standard quantum-mechanical notation. Suppose one has the commutator relation

$$AB - BA = C.$$

Take its matrix element between the states $\langle f|$ and $|i\rangle$, to give

$$\langle f|(AB - BA)|i\rangle = \langle f|C|i\rangle,$$

and insert the familiar resolution of unity

$$\sum_n |n\rangle\langle n| = 1$$

on the left-hand side, to give

$$\sum_n \{ \langle f|A|n\rangle\langle n|B|i\rangle - \langle f|B|n\rangle\langle n|A|i\rangle = \langle f|C|i\rangle.$$

This relation is called a sum rule. The final summation over n is usually, in an actual physical application, found to be expressible as an integral.

Appendix 3   Veneziano Model

The gamma function $\Gamma(x)$ has the properties:

(a) poles at $x = 0, -1, -2, \ldots$;
(b) $x\Gamma(x) = \Gamma(x+1)$; and

(c) $\dfrac{\Gamma(x)}{\Gamma(x-a)} \sim x^a$,   as $x \to \infty$.

Veneziano told us to consider the expression

$$V(s,t) = \frac{\Gamma(-\alpha(s)).\Gamma(-\alpha(t))}{\Gamma(-\alpha(s)-\alpha(t))},$$

where $\alpha(x) = \alpha'x + \alpha_0$, and α' and α_0 are constants.

Property (a) implies that V has poles when $\alpha(s) = n$ (a non-negative integer). Property (b), together with the linear form of α, can then be used to show that the residue at the pole is a polynomial in t of degree n — which corresponds to the presence of states with angular momenta up to $l = n$. Property (c) implies that

$$V \sim t^{\alpha(s)}, \text{ as } t \to \infty.$$

178

In other words, V has the classic properties corresponding to a family of Regge poles associated with $\alpha(s)$. But V is symmetrical in s and t and so it must *also* have the properties corresponding to a family of Regge poles associated with $\alpha(t)$. The simultaneous possession of these properties is precisely duality.

Two points are worthy of note. One is that the argument depends upon the linear form of the trajectory function α. In other words, the Veneziano model corresponds to the (empirically favoured) case of straight-line Regge trajectories. The other point is the multiplicity of angular-momentum states at $\alpha = n$. (This in fact implies that there is a whole sequence of parallel Regge trajectories involved; the lower ones are called 'daughters'.) Here we see something of the massive degeneracy of states involved in dual models.

Notes

The Proceedings of the *N*th 'Rochester' Conference are referred to as C*N*.

0 Pre-Rochester

1. Quoted in Pais (1986), p. 19.
2. See, for example, Polkinghorne (1984).
3. Wearing my clerical hat, I would claim it is an enterprise with theological justification. To understand the pattern of creation − at once rational (because of the reason of God) and contingent (because of his freedom) − is part of the search for the Logos. It is no accident that modern science began in the Christian setting of sixteenth-century Western Europe (see, for example, Russell (1985)).
4. This is due to the fact that the effective energy of collisions is that in the centre of mass of the beam and target particles.
5. See, for example, Polkinghorne (1979) for an account.
6. See any standard textbook on quantum mechanics, for example Schiff (1968) or Sudbery (1986).
7. See Pais (1986), pp. 313−20.
8. See, for example, Polkinghorne (1979) for details or, for a more technical account, J. Mehra in Taylor (1987), pp. 63−75.
9. Feynman (1985), pp. 7−8.
10. Modern superstring theories may show just this happening.
11. Pais (1986), p. 452.
12. This is a natural possibility in quantum theory because of the superposition principle; see Polkinghorne (1984), Ch. 4.
13. In Brown and Hoddeson (1983), p. 343.
14. *Ibid*. The account of New Testament theology is highly questionable!
15. In Brown and Hoddeson (1983), p. 382.
16. Pickering (1984), p. 10; italics in the original.
17. Quoted in Pais (1986), p. 8.

1 Rochester 1

1. R.E. Marshak, *Bulletin of Atomic Scientists* XXVI, 6 (1970).
2. Brown and Hoddeson (1983), p. 46.

3. *Ibid.*, p. 55.
4. Pais (1986), p. 461.
5. *Ibid.*, p. 480.

2 Rochester 2

1. Peierls (1985).
2. C2, Preface.
3. C2, p. 16.
4. Pais (1986), p. 495.
5. C2, p. 38.
6. C2, p. 33.
7. Pais (1986), pp. 511–2.
8. *Ibid.*, p. 513.
9. C2, p. 50.
10. C2, p. 56.
11. C2, p. 92.
12. C2, p. 93.
13. Polkinghorne (1979), p. 2.
14. C2, Appendix: 'Some People Don't Know When to Stop'.

3 Rochester 3

1. C3, Foreword.
2. *Ibid.*
3. C3, p. 5.
4. C3, p. 6.
5. C3, pp. 7–8.
6. See a standard textbook on quantum mechanics, for example, Schiff (1968). The phase shift is the angular difference in phase between the incoming and outgoing waves in that particular state.
7. The labelling is $(2I,2J)$.
8. C3, p. 29.
9. C3, p. 61.
10. Pais (1986), p. 488.
11. C3, p. 53.
12. *Ibid.*
13. C3, p. 47. (In Europe, as far as emulsions are concerned, there is Bristol, the great sun, and then a small number of little satellites whose dimension, even when added together, is very much less than that of Bristol.)
14. C3, p. 70.
15. C3, p. 73.
16. C3, p. 74.

4   Rochester 4

1. C4, Foreword.
2. C4, p. 20.
3. C4, p. 32.
4. C4, p. 92.
5. I usually use the notation Gev (originally European and now worldwide), but out of deference to American initiative I employ their original notation for these first results from a machine operating in that energy range.
6. C4, p. 91.
7. C4, p. 45.
8. *Physics Today*, January 1954, pp. 14f.
9. *Ibid.*, p. 17.
10. *Ibid.*

5   Rochester 5

1. Ref 1 of Ch. 1, p. 94.
2. *Ibid.*
3. C5, p. 20.
4. C5, p. 4.
5. C5, p. 38.
6. Quoted in Pais (1986), p. 516.
7. C5, p. 180.
8. Pais (1986), p. 492.
9. C5, p. 132.
10. C5, p. 134.
11. C5, p. 135.
12. C5, p. 136.
13. Quoted in Pais (1986), p. 516.
14. C5, p. 141.
15. M. Gell-Mann, 'Progress in Elementary Particle Physics, 1950-64', Caltech preprint CALT-68-1426, p. 18.
16. C5, p. 184.
17. *Ibid.*

6   Rochester 6

1. C6, p. VI-1.
2. C6, p. VII-10.
3. C6, p. V-7.
4. C6, p. VIII-1.
5. C6, p. VIII-7.
6. C6, pp. VIII-7−8.
7. C6, p. VIII-8.
8. C6, p. VIII-20.
9. C6, p. VIII-22.

10. C6, p. VIII-27.
11. C6, p. VIII-29.
12. Pais (1986), p. 525.
13. C6, p. I-10.
14. Ref. 15 of Ch. 5, p. 5.
15. C6, p. III-4.
16. C6, p. VI-28.
17. C6, p. IX-6.
18. C6, p. VIII-29.

7 Rochester 7

1. Ref. 15 of Ch. 5, p. 4.
2. C7, p. VII-1.
3. C7, p. IV-27.
4. It involved the extension of analytic ideas to multiparticle processes, a problem that Gell-Mann had suggested. I had invented a slightly inelegantly expressed version of what later were called generalised retarded products.
5. C7, p. IX-27.
6. C7, p. I-27.
7. C7, p. I-28.
8. The energy-dependence of a state of energy E is e^{-iEt}, so that if E is given a negative imaginary part, $-i\mu$, this produces an exponentially decaying factor, $e^{-\mu t}$, corresponding to a lifetime μ^{-1}.
9. C7, p. III-59.
10. C7, p. IX-11.
11. C7, p. IX-15.
12. C7, p. IX-27.
13. C7, p. IX-17.
14. C7, p. IX-7; my italics.
15. C7, p. IX-16.
16. See, for example, Polkinghorne (1984), Ch. 3.
17. C7, p. VI-1.

8 Geneva 8

1. C8, p. 39.
2. C8, *passim*.
3. *Physics Today*, November 1958, p. 23.
4. *Ibid.*, p. 20.
5. For fuller explanation See, for example, Marshak, Riazzudin and Ryan (1969).
6. This follows from a simple argument using angular momentum conservation.
7. C8, p. 291.
8. In DeTar, Finkelstein and Tan (1985), p. 23.
9. C8, p. 93.
10. *Ibid.*
11. C8, p. 94.
12. C8, p. 95.

13. Technical note. There is also a third variable, $u = (p_1 + p_4)^2 = (p_2 + p_3)^2$, corresponding to the channel $1 + \bar{4} \to \bar{2} + 3$. The variable u is linearly related to s and t: $s + t + u = \Sigma\, p_i^2$. Consequently, there are three terms in the most general Mandelstam representation, corresponding to dispersing in s, t; s, u; t, u; respectively.

14. C8, p. 96.

15. C8, p. 212.

16. This followed from the fact that analyticity prescribed that the imaginary part of the forward scattering amplitude behaved in the same way at infinity, however infinity was approached. Approach along the positive real axis gave the particle total cross-section (via the optical theorem); approach along the negative real axis gave the antiparticle cross-section (by crossing).

17. C8, p. 116.

18. C8, pp. 119ff.

19. C8, p. 75.

9 Kiev 9

1. *Physics Today*, December 1959, p. 40.
2. C9, p. I-473.
3. Ref. 1, p. 36.
4. C9, p. I-313.
5. C9, p. I-314.
6. Ref. 1, p. 36.
7. See, for example, Feynman (1985).
8. C9, p. I-6.
9. C9, p. II-99.
10. C9, p. II-77.
11. C9, p. I-327.
12. C9, p. I-443.
13. C9, p. II-122.
14. C9, p. II-309.
15. C9, p. II-296.
16. C9, p. II-411.
17. C9, p. II-412.

10 Rochester 10

1. *Physics Today*, December 1960, p. 20.
2. *Ibid.*
3. C10, p. 278.
4. Ref. 1, p. 22.
5. C10, p. 502.
6. C10, p. 508.
7. C10, p. 732.
8. C10, p. 858.
9. C10, p. 726.

10. C10, p. 806.
11. C10, p. 871.

11   Geneva 11

1. *Physics Today*, November 1962, p. 17.
2. Ed McMillan (C11, p. 724) pointed out that the consistent application of this principle would make the deuteron more fundamental than the neutron!
3. C11, p. 918.
4. See, for example, Ryder (1975), p. 198.
5. C11, p. 805.
6. These comments relate to orbital angular momentum. The theory can be modified appropriately to take spin into account, including, of course, the possibility of physical fractional values (½ ,3/2, etc.).
7. C11, p. 533.
8. Because of positive signature (Appendix 1) this does not give a particle pole.
9. C11, p. 533.
10. Physically they would be expected to correspond to shadowing corrections, which decrease the cross-section, but their sign was such as to augment it.
11. C11, p. 525.
12. *Ibid.*
13. C11, p. 530.
14. Ref. 1, p. 18.
15. Pais (1986), p. 571.
16. C11, p. 638.
17. C11, p. 854.
18. C11, p. 929.
19. C11, p. 930.

12   Dubna 12

1. N. Samios, *Physics Today*, December 1964, p. 38.
2. C12, p. 2-213.
3. See, for example, Davies (1984), Ch. 11.
4. C12, p. 1-809.
5. There is a long-running subplot in the theory of the time seeking to relate constituent and current quarks. It was not a very successful activity.
6. Polkinghorne (1979), p. 65.
7. Quoted in Pais (1986), p. 558.
8. C12, p. 1-834.
9. See Pais (1986), p. 559, for a detailed story of this episode.
10. This is hard to explain in terms other than mathematical. The generators of $SU(3) \times SU(2)$ are *either* $SU(3)$ *or* $SU(2)$ operators; the generators of $SU(6)$ include those which are products of these factors.
11. One can perhaps get the flavour of what is going on by remembering that boosts only look like rotations if one replaces time (t) by imaginary time (it). That factor of i is obviously dangerous.

12. C12, p. 1-427.
13. C12, p. 2-39.
14. Particle physicists instantly recognised the truthfulness of Watson's (1968) account of the competition to unravel the structure of DNA.
15. Ref. 1, p. 38.

13. Berkeley 13

1. C13, p. 3.
2. *Ibid.*
3. *Ibid.*
4. *Ibid.*
5. C13, p. 5.
6. *Ibid.*
7. *Ibid.*
8. C13, p. 8.
9. C13, p. 68.
10. C13, p. 79.
11. C13, p. 9.
12. This was a currently fashionable idea by which one could have up to n particles in the same state (cf. fermi statistics with $n = 1$ and bose statistics with $n = \infty$).
13. C13, p. 9.
14. That is to say, a frame in which they are moving very rapidly with respect to the frame in which the (equal-time) commutators are evaluated.
15. C13, p. 745.
16. Polanyi (1962).
17. C13, p. 103.

14 Vienna 14

1. Technical note; the partial-wave properties of a Regge pole contribution gave phase-shift behaviours corresponding to direct channel resonances, for the case of linear trajectories.
2. C14, p. 215.
3. C14, p. 91.
4. C14, p. 93.
5. C14, p. 253.
6. C14, p. 276.
7. C14, p. 37.
8. *Physics Today*, January 1969, p. 123.
9. Quoted in Pickering (1984), p. 172.

15 Kiev 15

1. C15, p. 95.
2. C15, p. 142.

186

3. C15, p. 144.
4. C15, p. 219.
5. C15, p. 248.
6. C15, p. 594.
7. C15, p. 437.
8. C15, p. 262.

16 Chicago 16

1. C16, p. 3-74.
2. C16, p. 3-76.
3. *Ibid.*
4. C16, p. 3-78.
5. C16, p. 3-79.
6. C16, p. 3-102.
7. C16, p. 3-45.
8. C16, p. 1-201.
9. C16, p. 1-488.
10. C16, p. 1-415.
11. C16, p. 2-169.
12. C16, p. 2-136.
13. C16, p. 2-139.
14. C16, p. 2-161.
15. C16, p. 4-315.
16. C16, p. 4-249.
17. C16, p. 4-251.
18. Josephson submitted this work in the Prize (= Research) Fellowship competition at Trinity College, Cambridge. I was the elector whose job it was to present his case. At our first meeting I had the most eulogistic referees' reports to read out. Before our second and final meeting, John Bardeen (the Grand Old Man of this sort of physics) published a *Physical Review Letter* in which he alleged that Josephson had made a mistake. The subject was too far from my own field for me to attempt to evaluate the situation. I told the other electors that if Josephson had made a mistake I was sure that it was much more interesting and creative than the correct work of most of us. He was elected. A few weeks later Bardeen withdrew his objections. A few years later Brian Josephson won a Nobel Prize for this work.
19. Quoted in Pickering (1984), p. 180.
20. The condition is $\Sigma\, Q_i = 0$, the sum of electric charges being taken over all the fermions involved.
21. C16, p. 2-299.
22. C16, p. 4-334.
23. C16, p. 4-336.
24. *Ibid.*; my italics.
25. C16, p. 4-340.

17 London 17

1. C17, p. IV-149.
2. C17, p. IV-171.
3. Pais (1986), p. 605.
4. Polkinghorne (1979), pp. 116–7.
5. C17, p. III-100.
6. C17, p. III-93.
7. C17, p. II-65.
8. C17, p. II-192.
9. C17, p. I-270.
10. C17, p. VI-2.
11. C17, p. VI-6.

18 Tibilisi 18

1. C18, I-Introduction.
2. C18, p. I-C134.
3. C17, p. I-B36.
4. C17, p. II-N111.
5. C17, p. II-N112.
6. C17, p. II-N113.
7. _Ibid._
8. C18, p. II-N111.
9. C18, p. II-N120.
10. _Ibid._
11. C18, p. II-T52.

19 Tokyo 19

1. C19, p. 771.
2. C19. p. 957.

20 Madison 20

1. Polkinghorne (1986), (1988).
2. C20, p. 1095.

21 What Was Happening?

1. Kuhn (1970).
2. Quoted in Newton-Smith (1981), p. 93.
3. Polanyi (1962).
4. In Leplin (1984), p. 140.
5. Quoted in Leplin (1984), p. 218.
6. A. Fine in Leplin (1984), p. 83.
7. Polkinghorne (1984), Ch. 8.
8. See Honner (1987); Murdoch (1987).
9. Newton-Smith (1981), p. 14.

10. Polkinghorne (1986), p. 22.
11. Kuhn (1970), p. 150.
12. Newton-Smith (1981), pp. 162ff.
13. Pickering (1984), p. 409.
14. Ch. 5, ref.15, p. 8.
15. Pickering (1984), p. 411.
16. In Leplin (1984), p. 230.
17. Clearly one is not concerned with trivially transient bad guesses like the dubnon.
18. Leplin (1984), p. 226.
19. In Leplin (1984), p. 145.
20. Newton-Smith (1981), p. 39.
21. See Newton-Smith (1981), Ch. VIII for a discussion.
22. Newton-Smith (1981), pp. 206ff.
23. In Leplin (1984), p. 29.
24. Leplin (1984), p. 22.
25. Barnes (1977), p. 24.
26. Pickering (1984), p. 406.
27. *Ibid.*, p. 413.
28. *Ibid.*
29. *Ibid.*, p. 409.
30. *Ibid.*, p. 410.
31. Feyerabend (1975).
32. Popper (1979), p. 319.
33. *Ibid.*
34. Popper (1963), p. 55; See also Popper (1980) − note the title!
35. Quoted in Newton-Smith (1981), pp. 67−8.
36. *Ibid.*
37. Pais (1982), p. 113.
38. For an account see Newton-Smith (1981), Ch. IV.
39. Quoted in Leplin (1984), p. 23.
40. R.N. Boyd in Leplin (1984), p. 60.
41. In Leplin (1984), p. 154.
42. *Ibid.*, p. 155; original in italics. See also Cartwright (1983).
43. *Ibid.*, p. 170.
44. Quoted in Thomson (1987), p. 8.
45. Newton-Smith (1981), p. 209.
46. Kuhn (1977), pp. 321−2.
47. Quoted in Longair (1984), p. 7.
48. Kuhn (1970).
49. Newton-Smith (1981), Ch. IX and XI.
50. Cartwright (1983), p. 6.
51. van Fraassen (1980).
52. In Leplin (1984), p. 243.
53. Newton-Smith (1981), pp. 208−9.
54. *Ibid.*, pp. 228−32.
55. Polanyi (1962).
56. *Ibid.*, p. 164.
57. Pickering (1984), p. 413.

Glossary

The definitions given are intended for the general reader. They do not attempt the precision that the professional would require.

analyticity   The powerful mathematical property of smooth behaviour in the complex plane (i.e. for values of energy, etc., extended to unphysical values, even including factors of i, the square root of -1).

angular momentum   A measure of the amount of rotational motion present in a system.

anomalous threshold   A singularity other than a *normal threshold* (q.v.).

anomaly   An extra term, unexpected on formal arguments but arising from the delicately singular nature of the theory.

asymptotic freedom   The property, possessed by gauge theories, that at short distances particles behave as if they were unconstrained by forces.

axial current   A current having a behaviour under reflection which is the opposite to that of the electromagnetic current.

bag model   A crude model of *confinement* (q.v.), which portrays quarks as retained within a rigid barrier (the bag).

baryon   A strongly interacting *fermion* (q.v.).

beta decay   A decay in which an electron is emitted.

boson   A particle of integral spin which, because of the *spin and statistics theorem* (q.v.), necessarily obeys bose (i.e. symmetrical) statistics.

channel   A particular scattering process. *Crossing* (q.v.) links together processes in different channels.

charge independence   The property that nuclear forces are independent of the charge on the nucleons involved. It is explained by *isospin* (q.v.).

charm   A quantum number of the *flavour* type (q.v.). It is associated with a particular kind of quark − the c.

colour   A quantum number which discriminates three varieties of each type of quark. It is also carried by gluons, but *confinement* (q.v.) implies that colour is a hidden quantum number, since all observable particles are 'white'.

commutator   The difference between two operators acting in different orders. The commutator of A and B is written (A, B) and is defined to be $AB - BA$.

complex plane   The set of numbers of the form $z = x + iy$, where i is the square root of -1 and x and y are (ordinary) real numbers. It can be thought of as forming a two-dimensional plane in which x and y are the coordinates.

190

confinement   The (hypothesised) property that quarks and gluons are permanently contained within the hadrons they compose and are never to be found existing on their own.

conserved vector current   The suggestion that the vector currents appearing in weak interactions are conserved currents (i.e. have properties similar to the electromagnetic current).

correspondence principle   The relationship by which a new theory approximates the results of its predecessor in the regime where the two theories overlap; in particular, the requirement that quantum theory reproduces the results of classical mechanics for large systems.

coupling constant   A measurement of the strength with which interactions take place which are due to a fundamental force of nature (for example the *fine-structure constant* (q.v.)). In modern physics, coupling constants are defined via residues at particle poles.

CP   The transformation which both interchanges right and left (P − reflection or parity) and interchanges particles and antiparticles (C − charge conjugation).

CPT   The transformation obtained by combining *CP* (q.v.) with T (time reversal, or the reversal of velocities). According to the *CPT theorem* all theories described by relativistic quantum mechanics are invariant under *CPT*.

crossing   The property that the function which describes the scattering process $1 + 2 \rightarrow 3 + 4$, it also describes the scattering process $1 + \bar{3} \rightarrow \bar{2} + 4$ (the crossed process in which an initial state particle is moved over to become a final state antiparticle, and vice versa).

cross-section   A measure of the effectiveness of the interaction between a projectile and the target. It is measured in terms of the area that the one appears to present to the other. The *total cross-section* refers to the effect of all that might happen; the *elastic cross-section* refers only to scattering in which the number and nature of the participating particles is unchanged.

current algebra   The mathematical structure specified by the *commutators* (q.v.) of components of currents (such as the electromagnetic current or the currents involved in weak interactions).

cuts   Most *singularities* (q.v.) are many-valued, that is to say, on encircling the singularity one returns to the same point but with a different value of the function. A cut is a mathematical device to produce a single-valued function by specifying that one is not allowed to complete the encirclement of the singularity by crossing the line of the cut. The difference in the value of the function just above the cut from its value just below the cut is called the discontinuity across the cut. It is simply the change in the function on encircling the singularity.

deep inelastic scattering   Scattering in which there is a large transfer of momentum to the target, which is broken up in the process. Such scattering penetrates the inner structure of the target.

degenerate   Degenerate states are states which have the same energy but differ from each other in other ways.

$|\Delta I| = \frac{1}{2}$ rule   The rule requiring that *isospin* (q.v.) changes in strangeness-changing weak interactions by half a unit.

diffraction peak   The behaviour of elastic scattering (i.e. scattering without any change in the participating particles) near the forward direction.

Dirac equation   The relativistic equation for the electron.

dispersion relations   Equations which hold for functions which are *analytic* (see *analyticity*) apart from certain simple singularities. Typically, the function is expressed as a certain integral of its discontinuity across the relevant *cuts* (q.v.).

Drell−Yan process   The production of a high-mass system through the fusion of a quark constituent of one incoming particle with an antiquark constituent from the other.

dual theory   A theory in which the *scattering amplitude* (q.v.) is built up by resonance contributions from any single channel and is not a sum of separate terms from different channels. The natural expression of duality is through *string theories* (q.v.).

eightfold way   A punning term for the SU(3) (see SU(*N*)) theory of hadrons in which eight baryons are grouped together.

~~**electroweak unification**   The successful combination of electromagnetic and weak~~ interactions into a single theory.

fermion   A particle of half-odd integral spin (1/2, 3/2, . . .). By the *spin and statistics theorem* (q.v.) such particles necessarily obey fermi (antisymmetric) statistics.

Feynman diagrams   Picturesque representations of terms in *perturbation theory* (q.v.) in which interactions take place through the exchange of *virtual particles* (q.v.).

Feynman integrals   Expressions for the contributions to the *scattering amplitude* (q.v.) associated with *Feynman diagrams* (q.v.).

fine-structure constant   The *coupling constant* (q.v.) of electromagnetism. Its value is approximately 1/137.

flavour   A facetious term for an explicitly observable hadronic *quantum number* (q.v.). Different flavours are associated with different quark types (u, d, s, c, b, . . .).

form factor   A measure of a hadron's interaction via a particular current (for example electromagnetic form factors). Essentially it signifies how 'spread out' the hadron appears from the point of view of that interaction.

gauge theory   The form a theory takes if it is to be invariant under transformations whose character can vary from point to point (gauge transformations). Necessarily gauge theories involve *vector particles* (q.v.) called *gauge bosons*. These theories are the modern forms of *quantum field theory* (q.v.) used in elementary particle physics.

ghost   A state which has associated with it the physically unacceptable property of a negative probability.

gluon   The *gauge bosons* (q.v.) of *colour* (q.v.) which cause the quarks to 'stick' together to form hadrons.

Goldstone boson   A massless boson which it was feared would necessarily occur with *spontaneous symmetry breaking* (q.v.). The Higgs mechanism (see *Higgs particle*) removes this difficulty.

graviton   The (so far hypothetical, but universally believed in) massless spin-2 particle which should be the carrier of gravity.

group theory   The theory of mathematical structures whose most important property is that they contain the analogue of multiplication. The groups of interest to physics mostly concern operations which are either space-time transformations (e.g. rotations) or shufflings of quantum numbers (for example interchanging particle types).

hadron   A particle with strong interactions.

Higgs particle   A hypothetical scalar (i.e. spin-0) particle thought to be a necessary ingredient of the *spontaneous symmetry breaking* (q.v.) in *gauge theories* (q.v.) that is required to given the gauge bosons mass. This process is called the *Higgs mechanism*.

192

hypercharge   The combination of *baryon* (q.v.) number and *strangeness* (q.v.) which appears naturally in SU(3) (see SU(*N*)).

hyperon   A *baryon* (q.v.) of non-zero *strangeness* (q.v.).

inclusive interactions   Experiments in which one final-state particle is monitored and the detail of the rest of what is happening is neglected.

infinity problem   The difficulty arising in *quantum field theories* (q.v.) when the calculation of quantities which should be finite yields nonsensically infinite answers. In certain theories it can be dealt with by *renormalisation* (q.v.).

infrared slavery   The hope that theories which are asymptotically free (see *asymptotic freedom*) at short distances will also produce such strong forces at large distances as to induce *confinement* q.v.).

intermediate vector boson   A particle of spin 1, acting as the intermediary of a fundamental force (particularly in weak interactions).

isospin   A quantity associated with the symmetry of strong interactions and expressing the close relationship between u and d quarks. (Originally called isotopic spin, it was first recognised in the discussion of the *charge independence* (q.v.) of the interactions of protons and neutrons).

jet   A group of energetic particles, all travelling in the same general direction, and often to be interpreted as the decay products of a primary entity (sometimes a confined quark or gluon).

Landau equations   The fundamental equations specifying the minimal *singularity* (q.v.) structure of a relativistic quantum theory (originally, the singularities of *Feynman integrals* (q.v.)). The locations of the singularities are said to lie on *Landau curves*.

lattice gauge theory   An attempt to calculate with a *gauge theory* (q.v.) by replacing space-time by a grid of discrete (i.e. separate) points.

lepton   A *fermion* (q.v.) without strong interactions.

Lie group   A *group* (see *group theory*) associated with a continuous set of transformations (i.e. transformations, like rotations, which can be made in as gradual a series of steps as one pleases). (It is pronounced 'Lee'.)

lightcone   A set of points in space-time which can all be connected to a single point (the vertex of the cone) by light signals.

low-energy theorem   Behaviour of a theory which will be found in the limit of very slowly moving particles.

magnetic moment   A measure of a particle's magnetic interaction.

Mandelstam representation   A hypothesised double-*dispersion relation* (q.v.), corresponding to treating both energy and momentum transfer as complex variables.

meson   A strongly interacting *boson* (q.v.).

muon   A charged *lepton* (q.v.), some 207 times more massive than the electron.

neutral current   A *weak interaction* (q.v.) in which no exchange of charge takes place.

neutrino   A massless *lepton* (q.v.). Neutrinos come in several varieties, corresponding to the varieties of massive lepton.

normal threshold   A particular *singularity* (q.v.), corresponding to the onset of a new physical process (e.g. the opening of a *channel* (q.v.) with an extra particle present).

nucleon   One of the constituents of nuclei, i.e. a generic term for protons and neutrons.

optical theorem   The relation between the total *cross-section* (q.v.) and the imaginary part of the forward-*scattering amplitude* (q.v.). It is a consequence of *unitarity* (q.v.).

p-wave   A state of angular momentum 1.

parity   The behaviour of a system under reflection.

parton model   A picture of hadrons as being composed of point-like constituents (partons).

PCAC   Partially conserved axial current, a smoothness property which enables *low-energy theorems* (q.v.) to be found for *pion* (q.v.) interactions.

peripheral interaction   One in which little momentum is transferred.

perturbation theory   A basic approximate calculational technique which seeks to expand quantities of interest in powers of a (small) parameter (usually a *coupling constant* (q.v.)).

phase   If a complex number $z = x + iy$ is written in the form $re^{i\theta}$, where r and θ are real, then θ is the phase. It may naturally be expressed as an angle.

phase shift   An angle (related to the *phase* (q.v.) of a *scattering amplitude* (q.v.)) which provides a natural parametrisation of what is happening in a particular angular momentum state.

phase space   A measure of the 'room' available for different final states in a scattering process. If the scattering follows phase space, it exhibits no special features.

pion   The lowest-mass *meson* (q.v.), the principal carrier of the long-range nuclear force.

polarisation effects   Phenomena depending on the *spins* (q.v.) − polarisation states − of the participating particles.

pole singularity   A particularly simple *singularity* (q.v.) corresponding to the behaviour of a/x near $x = 0$. The constant a is called the residue of the pole. In the *complex plane* (q.v.) of elementary particle physics, poles are interpreted as corresponding to single-particle intermediate states (*particle poles*), referring either to stable particles or to *resonances* (q.v.).

Pomeranchuk theorem   The result that particle and antiparticle total *cross-sections* (q.v.) have the same high-energy behaviour.

Pomeron   A particular *Regge pole* (q.v.) whose exchange gives high-energy behaviour corresponding to constant total *cross-sections* (q.v.).

potential theory   A non-relativistic way of describing interactions in terms of a potential energy which varies from point to point.

probability amplitude   Quantities which are natural objects of discourse in *quantum theory* (q.v.). They are complex numbers (i.e. of the form $x + iy$) and the probabilities are calculated from them by taking the square of their modulus (i.e. the quantity $x^2 + y^2$).

pseudoscalar theory   A theory involving particles of spin 0 and negative parity (pseudoscalar particles − usually pions).

quantum chromodynamics   The modern theory of *strong interactions* (q.v.). It is a *gauge theory* (q.v.) based on transformations of the *colour* (q.v.) degrees of freedom of *quarks* and *gluons* (qq.v.).

quantum electrodynamics   The *quantum field theory* (q.v.) of the electromagnetic interactions of *leptons* (q.v.); more specifically and orginally, the theory of the interaction of electrons and photons.

quantum field theory   A formalism in which quantum theory is applied to one or more

194

fields (quantities taking values at every point of space-time). All satisfactory formulations of relativistic quantum mechanics are quantum field theories.

quantum number   A distinguishing quantity associated with specific particles and conserved in at least some of the particles' kinds of interaction (the total amount present neither increases nor decreases in the course of the interaction). Examples are *strangeness, charm*, etc. (qq.v.).

quantum theory   The principles of mechanics essential for the discussion of systems so small that Heisenberg's uncertainty principle must be taken into account.

quark   One of the basic constituents of hadronic (see *hadron*) matter. Quarks come in various *flavours* (q.v.) and three *colours* (q.v.). One of their most distinctive characteristics is that they carry fractional electric charge (2/3 or $-1/3$). (The word is pronounced either with a short or a long 'a'.)

real axis   The set of real numbers (i.e. ordinary numbers, not involving i, the square root of -1). It forms the x axis in the *complex plane* (q.v.).

Regge pole   A *pole singularity* (q.v.) in the complex angular momentum plane of *Regge theory* (q.v.).

Regge theory   The theory based on treating the angular momentum as a variable in the *complex plane* (q.v.). It forms the basis for the discussion of behaviour at high energies.

relativistic quantum theory   The result of combining quantum theory with the requirements of special relativity. The union has deep consequences, e.g. the *CPT theorem* and *spin and statistics* (qq.v.).

renormalisation   The necessary act of replacing the sum of all contributions to a particle's mass and *coupling constants* (q.v.) by their physical values. If a theory is *renormalisable* this action then makes the theory's predictions finite.

renormalisation group   A mathematical expression for the way in which the effective strength of an interaction depends upon the distance over which it is sampled.

representation   A reproduction of the multiplicative structure of a group (see *group theory*) by means of 'shufflings' of a set of entities; also the entities so shuffled.

resonance   A short-lived particle-like state. The appearance of a resonance as a metastable intermediate state in a scattering process enhances the resulting *cross-section* (q.v.), hence the name. [A note on nomenclature: A particularly historically important resonance was the (3,3) (see p. 36), pronounced 'three-three'. Singly strange *hyperon* (q.v.) resonances are denoted generically by Y, doubly strange by Ξ, pronounced 'cascade' or 'xi' according to taste. Excited-state resonances (i.e. higher-energy ones) are usually given the superscript *; pronounced 'star'.]

S-matrix   The set of *scattering amplitudes* (q.v.) connecting all possible initial and final states in a scattering process. The *S-matrix approach* claimed that this was the only sensible entity to consider in seeking to formulate a theory of strong interactions.

scaling law   Behaviour in some high-energy regime where quantities of interest behave like a power of the energy multiplied by a function of dimensionless variables (i.e. ratios of variables measurable in the same physical units, such as the ratio of two energies or the ratio of an energy and a momentum transfer).

scattering amplitude   The natural quantities calculated in *quantum theory* (q.v.) are *probability amplitudes* (q.v.). Those associated with a scattering process (in which two particles collide, producing a particular final state (perhaps involving more or different particles) are called scattering amplitudes.

selection rule   The constraint imposed on the possible outcome of a process by the need to conserve some quantity (e.g. a charged particle cannot decay into purely neutral particles because electric charge is conserved).

singularity   A point in the *complex plane* (q.v.) where *analyticity* (q.v.) breaks down (i.e. a point of bad behaviour).

spin   The intrinsic angular momentum carried by a particle, independently of any angular momentum deriving from its state of motion.

spin and statistics theorem   A result of *relativistic quantum theory* (q.v.), that all particles with half-odd integral spin (1/2, 3/2, ...) must obey Fermi statistics (the wavefunction of a collection of these particles must be antisymmetrical under the interchange of any two of them) and all particles of integer spin (0, 1, 2, ...) must obey bose statistics (their wavefunction must be symmetrical under particle interchange).

spinor   In *relativistic quantum theory* (q.v.), the wavefunction of a particle of spin ½. In some sense this is the basic entity since other angular momentum states can be built up by adding and subtracting angular momentum ½. More generally, the term is used of the basic representation (in a similar building-brick sense) of any *Lie group* (q.v.).

spontaneous symmetry breaking   The property that the solutions of a theory can have a lesser symmetry than the theory itself. The precise detail of the way the higher symmetry is broken arises from the effect of infinitesimal and unpredictable triggers (spontaneously).

strangeness   A *flavour* (q.v.) quantum number, associated with presence of the s quark.

string theory   A version of relativistic quantum theory based, not upon point-like behaviour, but upon basic one-dimensional entities (strings) which trace out a two-dimensional sheet in spacetime.

strong interactions   Originally the strong forces which hold nuclei together. These and similar forces are now understood as manifestions of the underlying strong force of *quantum chromodynamics* (q.v.), binding quarks and gluons together in hadrons.

SU(N)   A *Lie group* (q.v.) involving a special kind of shuffling of N entities. (SU stands for special unimodular, the kind of shuffling which is natural in quantum theory.) The original *SU(3)* theory was the result of embedding *isospin* and *strangeness* (qq.v.) in a higher-symmetry theory.

supersymmetry   A high-order symmetry whose transformations include those exchanging *bosons* and *fermions* (qq.v.).

tau/theta puzzle   The perplexity arising when it was thought that there were two distinct particles, τ and θ, of opposite parities but coincident masses and lifetimes.

unitarity   The fundamental property that the probabilities of all the possible outcomes of a scattering experiment must individually be positive and together add up to 1 (something must happen).

V-particle   A neutral particle identified originally in cosmic rays by the V-forked track resulting from its decay into two charged particles.

$V-A$ theory   A theory of *weak interactions* (q.v.), involving a balance of *vector* and *axial currents* (qq.v.).

vacuum fluctuations   Transient blips of energy (the appearance and disappearance of *virtual particles* (q.v.)) present in the vacuum due to *zero-point motion* (q.v.).

vector current   A current whose behaviour under reflection is the same as that of the electromagnetic current.

196

Glossary

vector particle   A particle of spin 1.
virtual particles   Entities corresponding to internal lines in *Feynman diagrams* (q.v.), which do not have the mass value of a true physical particle.

weak interaction   The comparatively feeble force responsible for many particle decays.

zero-point motion   The inescapable quiveriness which Heisenberg's uncertainty principle imposes upon quantum systems, even in their lowest-energy states (the vacuum).

Bibliography

A 'Rochester' Conference Proceedings

Listed are: the general subject of the Conference; editors; publication details.

C1. No Proceedings are publicly available for the first Conference.
C2. Meson physics. A.M.L. Messiah and H.P. Noyes. Physics Department, University of Rochester (duplicated; 'Individual copies supplied upon request as long as the supply lasts').
C3. High-energy nuclear physics. H.P. Noyes, M. Camne, and W.D. Walker. University of Rochester and Interscience Publishers.
C4. High-energy nuclear physics. H.P. Noyes, E.M. Hafner, J. Klarman, and A.E. Woodruff. University of Rochester and the National Science Foundation.
C5. High-energy nuclear physics. H.P. Noyes, E.M. Hafner, G. Yekutele, and B.J. Raz. University of Rochester.
C6. High-energy nuclear physics. Prepared at the Princeton Institute for Advanced Study in collaboration with Princeton University and Brookhaven National Laboratory (six compiler-editors). Interscience Publishers.
C7. High-energy nuclear physics. Prepared at the University of Wisconsin by the Midwestern Universities Research Association (seven compiler-editors). Interscience Publishers.
C8. High-energy physics. B. Ferretti. CERN.
C9. High-energy physics. Academy of Sciences of the USSR, Moscow (2 Vols.).
C10. High-energy physics. E.C.G. Sudarshan, J.H. Tinlot, and A.C. Melissinos. University of Rochester and Interscience Publishers.
C11. High-energy physics. J. Prentki. CERN.
C12. High-energy physics. Ya.A. Smorodinsky. Atomizdat, Moscow (2 Vols.).(A complete version in English is also available: Israel Program for Scientific Translation, Jerusalem (1970).)
C13. High-energy physics. Margaret Alston-Garnjost. University of California Press.
C14. High-energy physics. J. Prentki and J. Steinberger. CERN.
C15. High-energy physics. V. Shelest. "Naukova Dumka" Publishers, Kiev.
C16. High-energy physics. J.D. Jackson and A. Roberts. National Accelerator Laboratory (4 Vols.).
C17. High-energy physics. J.R. Smith. SRC (Rutherford Laboratory).
C18. High-energy physics. N.N. Bogolyubov et al. Dubna (2 Vols.).

C19. High-energy physics. S. Homma, M. Kawaguchi, and H. Miyazawa. Physical Society of Japan.
C20. High-energy physics. Loyal Durand and Lee G. Pondrom. American Institute of Physics.

B. Other

Barnes, B. (1977) *Interests and the Growth of Knowledge*, Routledge and Kegan Paul.
Brown, L.M. and Hoddeson, L. (eds.) (1983) *The Birth of Particle Physics*, Cambridge University Press.
Cartwright, N. (1983) *How the Laws of Physics Lie*, Oxford University Press
Davies, P. (1984) *Superforce*, Allen and Unwin.
DeTar, C., Finkelstein, J. and Tan, C.-I. (eds.) (1985) *A Passion for Physics*, World Science Publishing.
Feyerabend, P. (1975) *Against Method*, Verso.
Feynman, R.P. (1985) *QED*, Princeton University Press.
Honner, J. (1987) *The Description of Nature*, Oxford University Press.
Kuhn, T. (1970) *The Structure of Scientific Revolutions*, Chicago University Press.
Kuhn, T. (1977) *The Essential Tension*, Chicago University Press.
Leplin, J. (ed.) (1984) *Scientific Realism*, University of California Press.
Longair, M. (1984) *Theoretical Concepts in Physics*, Cambridge University Press.
Marshak, R.E., Riazzudin and Ryan, C.P. (1969) *Theory of Weak Interactions in Particle Physics*, Wiley—Interscience.
Murdoch, D. (1987) *Niels Bohr's Philosophy of Physics*, Cambridge University Press.
Newton-Smith, W.H. (1981) *The Rationality of Science*, Routledge and Kegan Paul.
Pais, A. (1982) *Subtle is the Lord* ..., Oxford University Press.
Pais, A. (1986) *Inward Bound*, Oxford University Press.
Peierls, R.E. (1985) *Bird of Passage*, Princeton University Press.
Pickering, A. (1984) *Constructing Quarks*, Edinburgh University Press.
Polanyi, M. (1962) *Personal Knowledge* (2nd edn), Routledge and Kegan Paul.
Polkinghorne, J.C. (1979) *The Particle Play*, W.H. Freeman.
Polkinghorne, J.C. (1984) *The Quantum World*, Longman.
Polkinghorne, J.C. (1986) *One World*, SPCK.
Polkinghorne, J.C. (1988) *Science and Creation*, SPCK.
Popper, K. (1963) *Conjectures and Refutations*, Routledge and Kegan Paul.
Popper, K. (1979) *Objective Knowledge* (revised edition), Oxford University Press.
Popper, K. (1980) *The Logic of Scientific Discovery* (revised edition), Hutchinson.
Ryder, L.H. (1975) *Elementary Particles and Symmetry*, Gordon and Breach.
Russell, C.A. (1985) *Cross-Currents*, IVP.
Schiff, L.I. (1968) *Quantum Mechanics* (3rd edn), McGraw—Hill.
Sudbery, A. (1986) *Quantum Mechanics and the Particles of Nature*, Cambridge University Press.
Taylor, J.G. (ed.) (1987) *Tributes to Paul Dirac*, Adam Hilger.
Thomson, A. (1987) *Tradition and Authority in Science and Theology*, Scottish Academic Press.
van Fraassen, B.C. (1980) *The Scientific Image*, Oxford University Press.
Watson, J.D. (1968) *The Double Helix*, Weidenfield and Nicolson.

Index